Palgrave Studies in Austrian Economics

This series provides insights to the key debates within Austrian economics. It presents advanced research that links current discussions on Austrian economics within its rich and influential intellectual tradition. Bringing together books from the leading researchers in this field, the varied the series aims to define the current scope of research within Austrian economics and shape directions of future analysis, with entries that offer both a foundation to the topic for those new to Austrian economics and fresh insights and debates for those who are well versed on the topic.

Michael Esfeld • Cristian Lopez

Restoring Science and the Rule of Law

palgrave
macmillan

Michael Esfeld
Department of Philosophy
University of Lausanne
Lausanne, Switzerland

Cristian Lopez
Department of Philosophy
University of Lausanne
Lausanne, Switzerland

ISSN 3005-0464 ISSN 3005-0472 (electronic)
Palgrave Studies in Austrian Economics
ISBN 978-3-031-71185-5 ISBN 978-3-031-71186-2 (eBook)
https://doi.org/10.1007/978-3-031-71186-2

This Palgrave Macmillan imprint is published by the registered company Springer Nature Switzerland AG.
The registered company address is: Gewerbestrasse 11, 6330 Cham, Switzerland

Acknowledgements

In the first place, we would like to thank David Dürr, Guido Hülsmann, Carlo Lottieri, Thorsten Polleit and Henrique Schneider as well as Philipp Bagus and David Howden for introducing us to Austrian libertarianism and, in particular, the latter two for including this book in the new series on *Austrian Economics*—although it is not on economics specifically, but on the intellectual foundations of a just, peaceful and prosperous society. For many discussions on these topics as well as comments on the present manuscript, we are furthermore grateful to the members of the philosophy of science group at the University of Lausanne, namely, Alin Cucu, Michael Heumann, Guillaume Köstner, Alban Ridet, Amine Rusi El Hassani, Christian Sachse and Virginia Soro as well as to our local colleagues Olivier Massin and Eric Verrecchia.

Cristian Lopez gratefully acknowledges financial support from the Swiss National Science Foundation for his contributions to this book within the research project "Parsimonious ontology and the relationship between science and persons" (grant no. 105212B_200464). Some of the material included here draws on the German book *Land ohne Mut. Eine Anleitung für die Rückkehr zu Wissenschaft und Rechtsordnung* by Michael Esfeld, which was published by Achgut Edition Berlin in 2023. We are grateful to Henryk Broder, Dirk Maxeiner and Fabian Nicolay for granting us these rights. Finally, we would like to thank Brendan George, the philosophy editor at Palgrave Macmillan, for his support and professional handling of the manuscript.

Contents

CHAPTER 1

Introduction

Both modern science and the modern constitutional state, seeking to implement the rule of law, arose as liberation from a political order that was legitimated by a state religion. Both employ reason as a means to limit the exercise of power. In science, one has to provide evidence and arguments. Science strives for objectivity: it aims to discover the facts in the world as they are independent of our political, moral or religious beliefs. Consequently, in science, it is only relevant what somebody has to say and completely irrelevant what their social status, race, religion, gender, sexual orientation etc. are. By the same token, the rule of law implements the same rights and obligations for every person, independently of their beliefs about what is right and valuable for their personal lives as well as independent of their social status, race, religion, gender, sexual orientation etc. Science and the rule of law thereby constitute the two pillars of modernity as shaped by the age of Enlightenment. In contrast to nearly all the other cultures in history, Enlightenment, originating from the Christian religion, achieved the abolishment of slavery and eventually the banning of all sorts of racism from public life. It set the cultural background that led to an enormous scientific, technological and economic progress, creating a standard of living and quality of life for all segments of the population of which earlier ages could not even dream.

However, paradoxically enough, the very success of science and the constitutional state provoked a hubris that today risks ending up in their

M. Esfeld, C. Lopez, *Restoring Science and the Rule of Law*, Palgrave Studies in Austrian Economics,
https://doi.org/10.1007/978-3-031-71186-2_1

self-destruction. In science, the hubris consists in transgressing the border between discovering facts and prescribing norms as expressed in the slogan "Follow the Science". In the constitutional state, the hubris consists in transgressing the border between safeguarding the right to self-determination of every person in protecting them against unwanted external interference in their course of life and protecting them against all sorts of risks that can arise in life, thus turning universal *negative* rights of defence against interference into *positive* rights to all sorts of entitlements as in the welfare state. The result of this nefarious development is that we are about to return to a pre-enlightenment condition with science becoming a sort of secular religion—*scientism*—that legitimises an ever-increasing state power which regulates every aspect of social and even private life: *welfare totalitarianism*.

But how is it possible that constitutional states progressively transform into totalitarian states? One of the key elements of this transformation is the following: people are put under the general suspicion of endangering others through their everyday ways of self-determining their lives. Whatever they do out of their own free will in taking responsibility for their lives and engaging in all sorts of voluntary social interactions may contribute, for instance, to spreading a dangerous virus, to causing an apocalyptic climate change or to hurting the feelings of minority groups whose members suffered from injustice in the past. There is then a pervasive suspicion that everybody is a potential danger to everybody else. People cleanse themselves from this general suspicion by submitting to a comprehensive regulation and surveillance of their social and private lives that is imposed on them by an alliance of scientists, politicians, media and business interests. What used to be basic rights to self-determination is replaced with privileges for conformity with a totalitarian regime of all-comprehensive social control. This is the development that has recently become manifest in the corona, the climate and the wokeness regimes. Far from being merely exaggerated reactions to real risks or injustices, these are regimes in the sense of comprehensive regulations of social and even private life that are imposed on the people through the coercive power of the state. They thereby constitute the hitherto culminating point of the self-destruction of science and the rule of law in our civilisation.

Like at the beginning of modernity, we hence face today the task of liberating us from an overarching state religion in the guise of scientism. However, if science and the rule of law led to a hubris that results in their self-destruction, it would be a mistake to simply return to modernity:

there has to be something wrong in the way in which the age of Enlightenment set out to establish its two pillars. We argue that the mistake lies in the very idea of conceiving the republican, constitutional state as the guarantor to implement the rule of law and to foster science. There is nothing wrong with modern natural science and its method of achieving objectivity through disciplined scepticism, abstracting from all subjective features. There is nothing wrong either with the rule of law as an impersonal, non-arbitrary guarantor to enforce the obligation of every person to respect the natural rights to self-determination of every other person. The mistake is to seek to realise these aims through a central state whose organs have unlimited power in virtue of their monopoly in law making, law enforcement and judiciary as well as through declaring a state of emergency that entitles them to explicitly suspend basic rights.

By contrast, it is commonly taken for granted that the open society in the sense of Karl Popper (1945), in which different ways of life, religions, worldviews etc. can peacefully live together and mutually enrich each other, requires the framework of the modern central state. Such a state also asserts the mentioned monopoly through the regulation—or even direct financing through compulsory levies—of science and education as well as large parts of the media and the economy. The rationale is basically that a central state protects the values of the open society since it would prevent private interests from dominating any of these areas. To put it in a different way, the monopoly exercised by the state organs is intended to guarantee the pluralism that is vital to the open society, allowing people to pursue their goals in life as they think fit.

However, the idea that the values of the open society are protected or guaranteed by central states is wrong-headed. The concentration of power in the organs of the modern constitutional state leads to the mentioned hubris that ends in its self-destruction through the abuse of that power. It is an illusion to think that in order to establish a stable rule of law, a central state is needed that organises the public sphere. This is what we call the *illusion of republicanism*. To assert its power, the republican state turns in the first place into a welfare state: it makes the people dependent on its services by seizing the monopoly in providing what the people would otherwise organise out of their own free initiative, namely the insurance against central risks in life (such as old age, poor health and unemployment), thus crowding out voluntary associations. But it does not stop there: to vindicate its power, the welfare state tends to assure an always-growing provision of state-funded services. It thereby risks sliding into *welfare totalitarianism* by driving the provision of protection to absurdity—up to protection against the spread of viruses, the change in the world's climate and the expression of views that may hurt someone else's feelings.

One of the central features to identify welfare totalitarianism is the imposition of completely absurd things that people would never do out of their own free initiative—for instance, depriving the elderly of their social contacts through lockdowns and forcing them to die in isolation, or vaccinating one's children out of "solidarity" with a vaccine of which it is known from the outset that it cannot stop the spread of the virus in question. Indeed, it is by imposing completely absurd things that authorities confirm their exploitation of power. The state organs do so in the name of an ideology of a common good that requires transforming the existing people into new, allegedly better human beings—that no longer spread viruses, that consume energy without an impact on the environment, that hold and express only views that please everybody etc. In this regard, modern science then plays a crucial role: the organs of the welfare state become dependent on scientists legitimising "social engineering" (political scientism), intellectuals that promote the corresponding narratives (intellectual postmodernism) and state-fed businesses that deliver the corresponding products (crony capitalism).

To confirm its power, the republican state must also rely on collectivist narratives. These can be implemented in different ways. For instance, it can be a narrative that highlights the values of the open society, leading to relatively well functioning constitutional, albeit welfare states as in the Western world until recently. This was precisely the narrative during the Cold War, which was centred on demarcating the West from the Soviet communist empire. But when that empire collapsed with the fall of the Berlin Wall in November 1989, this narrative lost force. The year 1989 was not "the end of history" (Fukuyama 1989, 1992) in that there was a definitive victory of the constitutional state.

Quite to the contrary, the disappearance of the contrast between the narrative of the Western world and actually existing socialism led to an intellectual vacuum that was gradually filled by new, specifically postmodern collectivist narratives that live on constructing fake, postfactual simulations of reality and that are employed by the state organs to cling on to and extend their power. These narratives hook on superficially to the open society. However, in fact, they abandon the rule of law as well as science and usher in a society that is closed under a sequence of totalitarian narratives. The new collectivist narratives thereby provide the justificatory mechanisms for the transition from the welfare state seizing the monopoly of insurance against central risks in life to welfare totalitarianism. This is our explanation of why what seemed to be a firmly established rule of law could fade away so easily in the corona, the climate and the wokeness regimes.

It would be a mistake to explain our current situation in terms of a global moral degradation that has hit the world in recent decades. It would also be a mistake to put the blame on perverse worldwide elites that conspire to control the world. People have various beliefs, interests, values, capacities and virtues. There will always be intellectuals that are prone to totalitarian ideas, fearmongers that hallucinate the end of the world and scientists that misconceive science as a secular religion and thereby fall prey to superstition. There will always be business interests that seek to profit from promoting these ideologies, as the pharmaceutical industry in the corona regime, the "renewable" energies industry in the climate regime and the ESG labelling in the wokeness regime. There will always be collusion between these groups. The issue is not to change people in order for them to comply with some ideal of virtue, or to avoid collusion. The issue is to prevent that any group can deploy coercion to succeed in its aims. In other words, the issue is to use reason as a means to limit the exercise of power in such a way that no ideology can be imposed upon others, be it under the name of the "common good", "The Science" or any religion.

The "great reset" (Schwab and Malleret 2020), the totalitarian plans of "pandemic preparedness" of the World Health Organization, to mention but two contemporary examples, would be doomed from the outset if they could not seek to employ the coercive organs of the existing states to impose their totalitarian rule upon the people. And once there are states that are endowed with the mentioned concentration of power, their organs tend to maintain and extend their power and thus tend to fall prey to such totalitarian ideologies, if there are not particular historical circumstances—such as the ideological contrast between the Western civilisation and the communist empire—that prevent them from doing so. The problem lies in the institutional structures and their incentives in which individuals act. That is why it was a mistake of the Enlightenment age to set out to implement the rule of law and to promote science through the creation of the modern central state with its concentration of power through the monopoly in law making, law enforcement and judiciary. It is therefore no solution today to seek to invoke the coercive power of that state to eradicate the corona, the climate and the wokeness regimes; for that very power is the root—and not the cure—of the problem.

To put it in a nutshell, our diagnosis is that the current trend towards a new totalitarianism is the result of the convergence of four elements:

(1) political scientism;
(2) intellectual postmodernism;
(3) welfare state;
(4) crony capitalism.

The first two are the intellectual superstructure (the *Überbau*, to use a Marxist term), whereas the latter two are the real-world substructure (the *Unterbau*). Accordingly, to break the trend towards a new totalitarianism, we have to remove both the *Unterbau* by dismantling the coercive apparatus of the state with its monopoly of law making, law enforcement and judiciary and the *Überbau* through a new Enlightenment that does not repeat the mistakes of the old, eighteenth-century Enlightenment.

All this requires understanding the motivations and the roots of these elements. To do so, we retrace the conceptual foundations of modern science and the rule of law from René Descartes and Thomas Hobbes to the Enlightenment age with Immanuel Kant at its centre and pursue them with the social-normative, linguistic turn in twentieth-century philosophy. To overcome the mistakes of the age of Enlightenment, we then rely on the resources of Austrian libertarianism from Karl Popper, Friedrich von Hayek and Ludwig von Mises to Murray Rothbard and Hans-Hermann Hoppe.

The next two chapters of the book are an analysis of modern science. We elaborate on how it can fell prey to scientism by going briefly through its history from the nineteenth century to our days with the corona regime and the climate regime, working out how we get to the current trend towards a new, specifically postmodern totalitarianism. Against this background, we then lay out how the ideas of Descartes, Kant, Popper, Hayek and Mises enable us to restore science today and to immunise it against scientism.

In Chaps. 4 and 5, we then turn to the rule of law. We establish how a dilemma that Hobbes poses in the *Leviathan* (1651) leads to the conclusion that the modern central, constitutional state with its monopoly of law making, law enforcement and judiciary is on the one hand inevitable to implement the rule of law. However, on the other hand, that very state slides through its unlimited power into the welfare state and, finally, into welfare totalitarianism and crony capitalism with scientism then becoming the ideology that legitimises that all-encompassing state. We show how Austrian libertarianism and in particular the normative turn from Mises to Rothbard provide the conceptual tools to overcome Hobbes' dilemma and to build a stable rule of law without the monopoly of the central state.

Finally, we set out concrete steps towards a new Enlightenment that can achieve the goal of resuming the path of modernity by restoring science and the rule of law. The future is open. Western civilisation made great achievements that were previously unimaginable such as the abandonment of slavery. Our civilisation has the potential to continue on this path. We wrote this book with the intention to make a small contribution to providing the intellectual tools as well as the encouragement to do so.

CHAPTER 2

Science Versus Scientism

2.1 From Plato and Aristotle to Descartes: The Foundations of Modern Science

In the introduction, we mentioned that the enormous success of modern science leads to a hubris that may well destroy it. This hubris consists in the transgression from science to scientism as enshrined in the slogan "Follow the Science" (with capitals) that is omnipresent in our societies these days. To begin, it is a normative statement: it urges us to form a certain kind of beliefs and act accordingly. Second, it refers to "The Science" as a monolithic corpus of knowledge that enjoys a high reputation. Third, it compels us to form such beliefs and act accordingly in the face of daunting challenges and dangers that humanity must cope with. The mere existence of the slogan and its partisan instrumentalisation have many worrisome consequences. First and foremost, it is anchored in the idea that humanity faces unprecedented challenges: a new deadly virus or a human-driven climate cataclysm. Second, it is also anchored in the idea of "The Science", which not only certifies that the dangers are real and pressing but also offers the solutions. "The Science" takes the form of an unquestionable, rock-solid consensus that is in the position to put the blame on who is responsible for the dangers and to prescribe what we must do to cope with them. Finally, the normative nature of the slogan makes dissensus highly costly—whoever dares to not follow "The Science"

M. Esfeld, C. Lopez, *Restoring Science and the Rule of Law*, Palgrave Studies in Austrian Economics,
https://doi.org/10.1007/978-3-031-71186-2_2

in this specific sense is morally and epistemically degraded, sharing a place next to white supremacists and flat-earthers.

Tradition has it that at the door of Plato's academy, the following phrase was engraved: "Let no one ignorant of geometry enter". Contemporary academies could well engrave at their door the phrase: "Let only followers of the Science enter". There are indeed many striking analogies between the Platonic conception of knowledge and society and contemporary scientism. In this chapter, we briefly rehearse these analogies. We thereby show how both modern science and scientism can be traced back to two diverging paths that philosophy and the quest for knowledge in general have taken since their beginnings in ancient Greece: the Platonic and the Socratic-Aristotelian one. While the former has largely provided the elements for scientism, the latter has paved the way to the scientific spirit. We briefly recall the roots of these two paths in antiquity and then lay out the hallmarks of modern science. After that, we show how modern science can slide into scientism and go briefly through the history of scientism and its political use from the nineteenth century to our times, arguing that with the corona, the climate and the wokeness regimes we are about to slide into a specifically postmodern form of political scientism today that we dub "actually existing postmodernism".

Before Plato entered the scene, Socrates was well known for wandering around questioning elites' knowledge of justice, the common good, science etc., in the spirit of practising epistemic modesty with respect to what one thinks one knows. Thus, Socrates claimed no knowledge for himself—except of knowing that he does not know anything. Nonetheless, his scepticism was only methodological: he did not question the possibility of knowledge. He merely insisted that all knowledge claims must be subjected to scrutiny in order to find out whether they are justified. It then turned out that more or less all the knowledge that members of the elite claimed for themselves did not stand up to this scrutiny.

Socrates' questioning of knowledge claims marks a crossroads that leads to two divergent paths. The one path is to accept that there are no identifiable persons or groups of persons who have verifiable knowledge of the common good; this is the path that can be traced back to Aristotle and which later leads to modern science as well as to the modern, constitutional state. The other way is to leave Socrates' scepticism behind and to maintain that there are indeed identifiable individuals who, if they undergo the proper training, acquire knowledge of the universal good. This is the path that Plato takes in his magnum opus *The Republic*.

For Plato, knowledge in general is knowledge of ideas, that is, of archetypes, ideal forms of the sensible, concrete things that participate in the ideas. The ideas express not only what things *are* but also what they *should* be. Sensible horses, for instance, are imperfect copies of the ideal horse (the idea of the horse), which sets the norm of what being a horse is and should be. Accordingly, Plato finds it difficult to admit ideas for banal things that have no value whatsoever, such as mud and dirt (see *Parmenides* 130 c-e). We have terms to describe such things, but the ideal mud or the ideal dirt are not a norm in the sense of how something should be and with which a value is associated.

The normative character of the ideas enables Plato to move seamlessly from ideas of everyday things such as horses to ideas of values. Accordingly, the hierarchy of ideas reaches its climax in the idea of the good. This is the absolute good. It includes what is good for each of us. The decisive step then consists in the assertion that there are identifiable persons who have consolidated knowledge of the idea of the good. This knowledge is indisputable: once one has it, no doubt is possible. According to Plato, the philosophers acquire this knowledge. However, philosophers do not form an elite by birth. Rather, they reach this knowledge as the end point of a long and arduous path of training their own intellect and character. In principle, this path is open to everyone. But only a few have the ability to follow it to the end, achieving knowledge of the good.

The well-known consequence of this stance is Plato's postulate of the philosopher-kings. Because the philosophers—and they alone—have the knowledge of the universal good, they are not only entitled to rule in the state but also have the duty to do so or to instruct the rulers. The experts (the philosophers) therefore not only have a privileged knowledge but also, by possessing this knowledge, have the moral qualification and obligation to implement the good in society. Since the idea of the universal good is that of the absolute good, namely the idea of the good as such, there can be no conflict between what is good for a particular person or particular group of people and what is good for society as a whole. Consequently, the power of the philosopher-kings is unlimited. It does not only concern public affairs. There is no private sphere that is in principle exempt from the orders of the philosopher-kings. They can also determine who may marry whom and when in order to ensure the optimal continuation of society in the transition from one generation to the next.

Plato's political philosophy is in the first place a *collectivism*. Individuals only come into consideration insofar as they are part of a collective, namely the society or state to which they belong. They are completely absorbed in the function that they perform in the collective. This function is assigned to them by the philosopher-kings. It results from the general good, and it is also the good for the respective individuals, taking for granted that their life is absorbed by their role in the collective. Thus, given the ontological and moral superiority of the community over the individual, the Platonic society is characterised by the suppression of dissidence and any form of individuality, by promoting the destruction of existing forms of life that do not accommodate to its vision of the common good and by a regimentation of a strict re-education of the right, absolute idea of the common good. As this collectivism is based on knowledge claims, it is *political scientism*.

Scientism is the doctrine that scientific knowledge also encompasses the human being and all aspects of our existence, including thought, action, norms and moral values. Taken to its extreme, scientism is Plato's view that experts ultimately attain knowledge of the absolute good. Scientism becomes political when it is formulated as the programme to govern state and society according to scientific knowledge. This happens when Plato's philosophers become rulers or instruct the rulers. The result is *totalitarianism*: there is no limit to political rule that legitimises itself through knowledge claims (political scientism), if these knowledge claims themselves have no limit (scientism).

Accordingly, Karl Popper marks Plato's state as the origin of political scientism and totalitarianism in Western thought in his masterpiece *The Open Society and Its Enemies* (1945, vol. I, chapters 6–9). The intellectual basis of that book goes back to the exchange on scientism between Friedrich von Hayek and Karl Popper since Hayek's invitation of Popper to the London School of Economics (LSE) in 1936 and, later, Hayek's initiative to get Popper appointed at the LSE following his accomplishment of the manuscript of that book.[1]

In contraposition to the Platonic path, Aristotle takes up Socrates' scrutiny of knowledge claims in that he maintains that there neither is an identifiable group of experts who with certainty know what the common good is nor are there settled knowledge claims that may escape from methodological scepticism or epistemic modesty. In contrast to the Platonic society where everything is top-down organised, without dissent

[1]For a brief overview, see https://www.lse.ac.uk/philosophy/history/1946-and-all-that/. See furthermore Popper's homage to Hayek in Popper (1997).

and discussion by the ruling elite, in the Aristotelian society as set out in the treatise *Politics*, citizens gather publicly to deliberate about the public affairs, without relying on, or obeying to, a selected group of experts. They take the decisions about public affairs in common and thus participate in these decisions. In a nutshell, the Socratic-Aristotelian path is one of open, horizontal deliberation among peers, where truth can only be reached after scrupulous scrutiny of knowledge claims, by the exercise of methodological scepticism and epistemic modesty.

The contrast between Plato, on the one hand, and Socrates and Aristotle, on the other, illustrates two tendencies in Western philosophy that pull our attitudes to knowledge towards two opposing directions since Ancient Greece. Many Platonic traits can be identified in contemporary scientism. The ironclad suppression of dissenting voices against alleged consensus and settled knowledge has been inflexibly encouraged in the corona regime and the climate-change narrative. Paradoxically, any trait of methodological scepticism or epistemic modesty became a trait of bad-intentioned rejection of alleged facts, anti-scientific spirit or plain moral degradation. Yet, these Socratic-Aristotelian features have laid out the fundamentals of modern science and the scientific spirit.

In 1637, René Descartes published his *Discourse on Method*. Along with Galileo Galilei's works, it is one of the turning points on the road to modern science. While few survives of Cartesian science per se, the way in which Descartes understood the scientific spirit and method has gone deep in modern science. In the sixth part of the *Discourse*, Descartes advances the famous statement that science should be employed to make us masters and owners of nature ("nous rendre comme maîtres et possesseurs de la nature", Descartes 1902, p. 62).

By "nature", Descartes here means first and foremost human nature. The goal is to develop medicine to fight diseases and epidemics, which had swept Europe for almost three centuries. For instance, the plague killed about 10–30% of the population in Europe in the seventeenth century. By contrast, the infection fatality rate of the coronavirus wave in spring 2020 was about 0.3–0.4% (e.g. see Ioannidis 2020b). Non-human nature then comes also into focus, namely, to gain knowledge about the laws of motion of matter in order to improve people's living conditions through technological progress. Again, in Descartes' time, about 95% of the workforce were employed in agriculture to secure the daily survival and only about 5% could pursue other activities. By contrast, in our times in Western societies, it is the other way round with only about 5% of the workforce being employed in agriculture.

In setting natural science's goal in terms of improving people's living conditions, Descartes develops a different understanding of scientific knowledge from that dominant in antiquity and the Middle Ages. Science is not the disinterested observation of nature—*theoria* in the literal Greek sense of contemplation—but serves a practical purpose. It is thereby directly linked to technology. However, in order to achieve that practical purpose, *objectivity has to be the hallmark of modern science*. All its further characteristic traits—as well as the contrast with scientism—follow from the strive for objectivity.

Objective knowledge of how non-human nature as constituted independently of our subjective and evaluative judgements is necessary in order to improve our living conditions through technological progress based on knowledge. By abstracting from all subjective features, Descartes gets to conceiving non-human nature as *res extensa*, characterised solely by extension and motion. If only extension and motion are objective, there is no place in modern science for Platonic Ideas or Aristotelian intrinsic forms. Through the strive for objectivity, modern science thereby sets a cleavage between facts and norms.

The ideal of objectivity is well expressed in terms of what is known as the "point of view from nowhere". This is the point that is reached by abstracting from all subjective elements of how things appear to a person (how they smell, look etc.). What then is left are only the relative positions of objects (extension) and their change (motion); for only these can be checked by scientific experiments (see Esfeld 2020, chapter 1.1 and 1.2). The strategy of formulating a kind of knowledge that is centreless and featureless (Nagel 1986, p. 14) is the basis for the success of modern science. This does not mean that modern science has ever achieved full objectivity, nor that it is even achievable. Strictly speaking, the idea of a point of view from nowhere is contradictory. A point of view from nowhere obviously is no point of view anymore, that is, no theory conceived in a language with a semantics and a pragmatics. The point of view from nowhere is a regulative idea that guides theory construction and empirical research. It serves as a standard against which claims to knowledge can be assessed. However, one can never completely abstract from all subjective features. Objectivity is important not because it is fully achievable, but because it makes science filter out as much as possible any trace of subjective influences.

Objectivity also means that anyone can understand scientific data collection and theory construction and undergo the appropriate training, regardless of social status, race, religion, gender, sexual orientation etc. There is no Chinese mathematics, physics, chemistry, biology etc., as opposed to Western or African ones. Nor is there a mathematics, physics, chemistry, biology etc. of Muslims, indigenous peoples, women, or transgender persons as distinct from the mathematics, physics, chemistry, biology etc. taught and researched in our educational institutions (so far, anyway). There is only one mathematics, physics, chemistry, biology etc. developed by the use of reason, which unites all human beings. It is hence misguided to regard scientific knowledge as "situated" or "multicultural" (Massimi 2022). Relativising knowledge to the place where people happened to be born, or to the sex that people happened to be born with is to place oneself out of the scientific spirit and the use of reason. That is how science employs reason as a means to limit the exercise of power: only argument and evidence count. Social status, race, religion, gender, sexual orientation etc. are insignificant when it comes to science.

In a trivial sense, all knowledge is situated since individuals produce scientific knowledge within societies and cultural environments. Scientific knowledge is an epistemic, collective enterprise of social collaboration. But the deliverances of scientific knowledge and their justification are independent of individuals' situations or their cultural environments. Some societies and cultures clearly foster scientific knowledge. However, it is a mistake to regard such a knowledge as the property of a specific society and culture. It is also a mistake to regard under-represented communities' knowledge as "alternative science"; insofar as they follow the scientific method, they equally participate in the elaboration of scientific knowledge.

In this sense, scientific knowledge is better regarded as *a*cultural or *a*perspectival than as intrinsically multicultural or perspectival. There is only one physics, chemistry, biology etc. because science seeks objectivity. Without striving for *one* science free from subjective influences, we leave the scientific realm. It is insightful to distinguish between a context of discovery and a context of justification (e.g. see Leplin 1987; Schickore 2014). While the context of discovery gives the details and contingencies that lead to the formulation of a theory, the context of justification gives the assessment of a theory's epistemic support (i.e. of a theory's validity according to universal standards). Making the validity of scientific findings dependent on social status, race, religion, gender or sexual orientation is not only a confusion between these two contexts but also an expression of an attitude that is hostile to the scientific spirit. If science depended on such subjective features, then modern science would not be possible; there would simply not have been the technological progress and improvement of our living conditions.

Finally, scientific objectivity means that scientists must strive for scientific integrity, in the sense of avoiding socially pervading biases (such as "social desirability biases", "conformity/unconformity biases" etc.). The striving for scientific integrity is to reflect the facts as unbiased and complete as it is possible, avoiding personal judgements, political interests or well-entrenched prejudices. In a nutshell, scientific objectivity encourages us to form beliefs according to theories supported by evidence and argument, and not to form beliefs along the lines of social status, race, religion, gender, sexual orientation etc.

Richard Feynman, for one, was a strong defender of scientific integrity in this sense. His lecture on "The Value of Science" (1955) has the following subtitle:

> Of all its many values, the greatest must be the freedom to doubt. (Feynman 1955, p. 13)

He then describes scientific knowledge in these terms:

> Scientific knowledge is a body of statements of varying degrees of certainty—some most unsure, some nearly sure, none *absolutely* certain. Now, we scientists are used to this, and we take it for granted that it is perfectly consistent to be unsure—that it is possible to live and *not* know. (Feynman 1955, p. 14)

This leads us to a second central feature of modern science: ***disciplined scepticism***. Indeed, it is uncertainty that moves knowledge forward and motivates the evolution of science. Absolute certainty, by contrast, is dogmatism, frequently justified in terms of authority—be it the authority of an individual, be it the authority of most scientists in a field at a time. Not only does authority play no role in science, but also it goes straight against it. The Socratic-Aristotelian path exercises methodological scepticism and is therefore opposed to authority-established knowledge.

Historical considerations show how short-lived scientific consensus has often been. To consider one example, in 1894, Albert Michelson said at the inauguration of the Ryerson Physics Laboratory at the University of Chicago that most of the grand underlying principles have been firmly established and that future physics is mainly about improving measurements in the sixth place of decimals (Michelson 1894, p. 159; see also Michelson 1903, p. 24). Max Planck (1924, p. 103) also reported that most physicists at the turn of the nineteenth century believed that theoretical physics was nearly to be finished. Yet, the developments of special and general relativity and quantum mechanics in the first decades of the twentieth century showed how much off-target such claims were. Modern science was not settled then, and it is not settled now.

It is important to note how harmful and anti-scientific the appeal to "The Science" as authority can be. During the first coronavirus waves in 2020, Lothar Wieler, the president of the German Robert Koch Institute, declared on 28 July 2020 that corona rules such as mask-wearing must not be questioned (see Ärtzeblatt 2020). This means to impede the exercise of methodological scepticism, to not try to scientifically scrutinise the claim that general mask-wearing is a suitable means to stop the spread of the virus. It just destroys modern science at its core. In a similar vein, in an interview with *Face the Nation* (CBC News, 28 November 2021), Anthony Fauci made this declaration:

> But if they get up and really aim their bullets at Tony Fauci, well people can recognize that there's a person there, so it's easy to criticize, but they're really criticizing science because I represent science. That's dangerous.

This attitude resembles the absolute monarch Louis XIV who, allegedly, said that "L'État, c'est moi!", in an attempt to block any questioning from the French Parliament.

Equally worrisome is the communication of partial or exaggerated scientific data to intentionally support a particular narrative. Scientists may thus think that they suffer from a "double ethical bind" (see Schneider 1989): when communicating scientific information, they must be honest, but also effective, in the sense of generating some specific response from the public, even if they thereby frighten the public with the possibility of an unprecedented cataclysm, or a dire emergency that will kill their beloved ones. This is clearly a mistake that not only undermines scientific objectivity but also clouds scientific integrity.

The idea that *disciplined scepticism* is the central pillar of the scientific methodology is already anchored in René Descartes' *Discourse on method.*

It is a direct consequence of the strive for objectivity being the hallmark of modern science. The point of view from nowhere is a regulative idea that can never be fully realised, because all scientific knowledge is generated within a specific historical and cultural context. No scientist can ever abstract from all subjective features in doing science. That is why disciplined scepticism is the characteristic feature of the scientific method: every claim to scientific knowledge (objectivity) has to undergo rigorous scrutiny, because it is conceived in a particular subjective context (which includes here the inter-subjective, historical and cultural context). And no claim to scientific knowledge can therefore be ever established as being objective beyond any possible doubt, being "*absolutely* certain" as Feynman puts it in the quotation above.

The sociologist Robert Merton (1942) characterises the institutionalised science of the twentieth century as "organised scepticism". The idea is simple: science is about the explanation of regularities. Any assertion about the explanation of such regularities only can stand up under critical scrutiny. Hypotheses are proposed to explain regularities. But the endorsement of a hypothesis is always negative: it is endorsed not because it has been shown to be true, but because there are no weighty reasons to deem it false. Therefore, the hypothesis is adopted only hypothetically, always trying to find something somewhere in the universe that contradicts it.

Consider a simple example: Let us assume that all observed ravens are black. On this basis, we formulate the hypothesis "All ravens are black". In order to establish this hypothesis as knowledge, it makes no sense to search everywhere for more black ravens—and find countless of them. On the contrary: this hypothesis is only confirmed if one does not succeed in finding ravens that are not black, despite making all feasible efforts to do so. This means that this hypothesis must be accepted only sceptically, being always subjected to find something somewhere in the world that can contradict it. If and only if any such attempt fails, the hypothesis is corroborated and tentatively endorsed as knowledge. Because the world can only be examined from a limited perspective, critical questioning is the way to move from what appears to be knowledge to a tentatively established knowledge.

Karl Popper highlights this procedure in his *Logic of scientific discovery* (1935): an assertion is a scientific statement if and only if one can precisely lay out under which circumstances it would be proven false and thus falsified. In this respect, the slogan "Anything goes" by Paul Feyerabend (1975) is also coherent: there are no taboos in science; no one has a privileged access to some unquestionable truth. Any scientific doctrine can always be challenged by pointing out new data or weighing arguments in a new manner.

This idea is also contained and enforced in the United Nations Economic and Social Council's general comment no. 25 (2020) when referring to scientific knowledge. Scientific knowledge is not based on authority or tradition, but on an intersubjective rational process open to everyone:

> Although protection and promotion as a cultural right may be claimed for other forms of knowledge, knowledge should be considered as science only if it is based on critical inquiry and is open to falsifiability and testability. (United Nations 2020, § 5)

Consequently, there is nothing like "The Science", the unperturbed monolith that policymakers and mass media so frequently rely on and many scientists like feeding. Modern science is intrinsically plural, in the sense that several, often dissonant voices coexist. It is a complex network of delivering and receiving reasons about facts: any voice that gives fact-and-argument-based reasons has a say in this network and is to be listened to. As regards science, the slogan "Follow the Science" is as empty as it is meaningless.

Finally, *modern science is intrinsically anarchic*. When it comes to science, no authority is required to determine what the corpus of scientific knowledge is. Majority vote or setting up a person or body to settle this—that is, to be the judge of scientific truth—would obviously be absurd and counterproductive to scientific progress. Only evidence and argument count. There is no rule required to establish anything as scientific knowledge and enforce it under threat and, if necessary, use of coercion. It is not a problem that competing hypotheses and theories often coexist in science. On the contrary, this is a motor of scientific progress. It would be contrary to the use of reason on which science is based to resolve disputes about claims to scientific knowledge by resorting to coercion, insofar as these disputes do not dissolve themselves through evidence and argument that are recognised by all parties of their own accord.

Scientific knowledge is without doubt a story of great success. The ideal of objectivity is at the centre of this story. But one should bear in mind that it is also a story of numerous mistakes, dead ends and false theories. Both stories are the two sides of the same coin: scientific knowledge is fallible, inherently uncertain. Our scientific knowledge is built upon provisory hypotheses that have *thus far* fared relatively well the tribunal of experience. But certainty is not the end of scientific knowledge. The recognition of uncertainty, that scientific knowledge can be (and will surely be) refuted by future evidence, is just to accept some degree of epistemic modesty. It is our best principle against dogmatism, and the best guarantee to respect other people's views. Scientific knowledge must adopt epistemic modesty not only because of uncertainty but also because of an environment in which a plurality of views can be freely exercised is the best-known way to progress. Epistemic modesty leads admitting other voices to be heard, to evidence to be freely scrutinised and to arguments and theories to be freely discussed. This, then, is the reason why scientific knowledge is a story of great success.

2.2 From Science to Scientism

Modern science as conceived by Descartes relies on the features we mentioned in the previous section. The idea of scientific objectivity abstracts from all subjective features including the ways in which we perceive the world through our senses, namely colours, sounds, smells, tastes and the like. Only the positions of things and their motions meet the benchmark of objectivity. Nonetheless, through the mathematical representation of positions and their evolution according to laws of motion, Cartesian science establishes a direct connection to the perceived objects and their change. As mentioned in the preceding section (and see Esfeld 2020, chapter 1.1 and 1.2), if a scientific theory correctly represents the spatial arrangement of things and its change, then it has correctly represented everything that can ever be intersubjectively verified in observations and scientific experiments. The natural science that goes back to Descartes is geometry, kinematics (the study of motion), and dynamics (the study of the forces that change motion). It has a direct relation to empirical data, including data gained through scientific experimentation.

Newton's law of gravity, for example, allows us to calculate the attraction of two bodies due to gravity. To do so, however, one must know the initial values of their position and velocity. If one sets out to calculate the trajectory with which a stone falls to earth when thrown, small uncertainties in the determination of these initial values are negligible. However, this is not the case if one throws a coin instead of a stone. The smallest variations in the initial values of positions and velocities can change the entire result of the coin toss—heads or tails. Therefore, one cannot predict how the coin will fall. One cannot measure the initial values of position and velocity so precisely that one can make a prediction of the outcome of the individual coin toss.

This is why Cartesian science—geometry, kinematics, dynamics—needs to be complemented by statistical mechanics, which was developed at the end of the nineteenth century. Statistical mechanics employs an abstract mathematical space to represent particle configurations and their evolution. By defining a probability measure on this mathematical space, one can make precise statements about the statistical distribution of measurement outcomes. Statistical mechanics is therefore the key to the success of applying physics to all real-life situations where we do not know the exact initial conditions.

The introduction of statistical reasoning into modern science strongly enhances it. However, one must be clear about the limits of this success: the case of the stone toss, in which one can make predictions of the individual outcomes with the help of deterministic laws of motion, is rather the exception. The case of the coin toss, in which one cannot make predictions of individual outcomes, is the generic situation. We have a useful tool to deal with this situation in statistical mechanics. Nevertheless, there is an evident limitation to scientific knowledge: while modern natural science can uncover general laws of nature, it is an illusion to believe that this science is therefore able to predict the future evolution of physical systems with certainty. If this is true even for simple physical systems, the limitation rapidly magnifies when the evolution of complex systems such as a virus wave or the world's climate are under the spotlight. Unfortunately, many scientists promote this illusion in their public communications. They thereby neglect the following truth: *The world may be knowable through natural science, in the sense of knowing the laws of nature. But this knowledge does not make it predictable and controllable.* This is one of the limits of modern science.

In this context, there is another achievement of natural science that is also prone to misuse: the employment of models. Models allow us to precisely play out possible scenarios of the evolution of the systems under consideration. However, in order to do so, one must first select the parameters that one considers relevant for the time evolution of the systems under consideration in the present circumstances. It is possible that parameters that are actually relevant are overlooked and parameters that actually do not make a significant difference in the situation at hand are assumed to be significant. For example, the model of the first coronavirus wave by Neil Ferguson et al. (2020) overlooks the crucial parameter of people's spontaneous behavioural adaptations. This leads to the misguided prediction of an unlimited exponential increase in infections unless the political authorities intervene with coercive measures. To mention another example, the existence of complex positive-negative feedbacks in the world's climate is also part of the parameters that may make a difference in model projections. We will consider these cases in Sects. 2.4 and 2.5.

Furthermore, model projections can widely vary depending on the exact initial values that are plugged in the model. Since uncertainty in the initial values leads to consider a range of possible values, models typically deliver a range of possible outcomes. This translates into different scenarios, compatible with the same models, but dependent on which initial value was chosen. This is not a defect of the scientific methodology as uncertainty is part of scientific endeavours. Irresponsibility and dishonesty arise when this leeway is employed to intentionally select some specific values to generate apocalyptic scenarios, justifying fearmongering and coercive measures.

One must therefore not confuse models with representations of reality. They are simulations. They are useful in that they run through possible scenarios of the future evolution of the systems under consideration. But they are not predictions of what time development these systems will actually take. At most, they can indicate statistical frequencies when dealing with a large number of identical initial situations, such as coin tosses. But waves of a new virus or the development of the world's climate are unique processes. Suggesting to the public that one can predict their future development by means of models is not science. It amounts to deceiving the public with the aim of using science as a pretext to deprive people of their basic rights of self-determining their lives. Models are undoubtedly scientifically useful. But there are, again, clear limits to what can be known and achieved by modelling. *The misuse of models therefore is the first step from science to scientism.*

Natural science is more than just physics and its many applications. Physics is universal in its fundamentals. It applies to all objects in the universe. It uncovers universal patterns of motion and formulates laws for them, such as the law of gravity. But during the evolution on Earth, special systems have also developed with novel, emergent properties, such as organisms. In the twentieth century, the study of these systems was successfully integrated into natural science, which is based on physics. The key to do so is functionalism. Starting from configurations of physical objects and their movements, described by the laws of physics, everything else can be defined by a function in the sense of the role that it plays in the movement of the objects in question. The study of high-level systems (as molecules, genes, organisms etc.) can thus be integrated into natural science based on physics and the corresponding scientific methodology (see Esfeld 2020, chapter 2, for details and references).

Let us look at a simple example: water. According to modern science, there is no specific water-substance. The ancient doctrine of the four elements of earth, water, air and fire has been superseded by modern science. But, of course, there is *water*: there is a substance that fulfils the functional role of being odourless, colourless, liquid at normal room temperature and quenching thirst through the changes it triggers when drank. Scientific research shows that this role is fulfilled by configurations of H_2O molecules and its isotopic variations. Thus, water is defined by its functional role that uses the vocabulary commonly employed to characterise water without reference to chemistry. But this functional role is related to its effects on physical movements. This then makes it possible to find configurations of physical objects that fulfil this role through scientific research—such as configurations of H_2O molecules in the case of water. In this way, the novel emergent properties that characterise water are integrated into the scientific description of the world based on physics.

Let us consider another example: living organisms. There is no evidence for an *élan vital* or specific vital substance, as the vitalists thought in the nineteenth century. But there are living organisms. Today's biology defines life functionally by characteristic patterns of movement such as reproduction and adaptation to the environment. Molecular biology finds out how these characteristics are realised by certain configurations of molecules. A milestone in this respect was the discovery of the molecular composition of the DNA by James Watson and Francis Crick in 1953. This discovery made it possible to show how the connections between genotype and phenotype, which classical genetics explores, are realised on the molecular level. Thereby, the way was paved to also integrate the theory of living organisms into the scientific description of the world based on physics.

Functionalism as a scientific method is in principle unlimited. In a sense, it is merely a matter of definitions. One can simply stipulate that anything that is not characterised in physical terms is to be described by its functional role, which is ultimately a role for the movement of matter and is therefore fulfilled by configurations of physical objects. In essence, functionalism is committed to the idea that the properties of some domain (e.g. special sciences or morality) are functional properties that are *realised* by physical properties. That is, the physical configuration of matter instantiates the property at issue.

Functionalism has become mainstream in philosophy since the 1970s and has different versions. One version is *role* functionalism (Fodor 1974; Putnam 1975), claiming that higher-order properties are role properties (playing a causal role), while physical properties are realisers. Role and realiser properties are not identical. Another version is *realiser* functionalism, according to which higher-order properties refer to realiser properties (physical properties) so that there is no ontological distinction between them (Lewis 1966; Armstrong 1968). Leaving these differences aside, our intention here is to stress that if everything is defined in functional terms, then the scientific method could well be unlimited, paving the way to scientism: every high-order property (as mental or even moral properties) can ultimately functionalised in terms of a physical basis within the range of science.

An important consequence then is that functionalism can also be applied to the human mind—the thoughts and actions of persons. Mental properties thus are to be defined by certain functional roles, which are ultimately roles for the physical behaviour of persons, and are, therefore, physically realised in the brain—just as the properties that characterise living organisms are realised by certain molecular configurations. Obviously, a person's thoughts, values, goals and intentions manifest themselves in behaviour. On this basis, one can even go so far as to treat ethics functionally as well: starting from the normative, moral attitudes that people have in fact, functional definitions of these attitudes can be formulated, which are ultimately definitions in terms of particular behavioural dispositions—that is, how people behave in particular situations given their moral attitudes (e.g. see Jackson 1998, chapters 5–6).

As previously mentioned, *the unrestricted application of functionalism promotes the transition from science to scientism*. Scientism is in the first place the idea that the range of science in the sense of modern natural science is unlimited. It includes everything, in particular also human thoughts and actions and thus also the social and the moral spheres. Hence, there is no principled distinction between science in the sense of the natural sciences and the humanities. Also, the humanities and social sciences, insofar as they are a rational enterprise, come under the paradigm of natural science set by physics, as do chemistry and biology. More precisely, against this background, *scientism is the claim that all meaningful questions, insofar as they admit of an answer by employing reason, can be answered by science in the sense of modern natural science, including the questions relating to human thought and action, in particular human freedom and free will, norms and values* (see most recently Peels 2023; see also the essays in Boudry and Pigliucci 2018).

Scientism, therefore, glosses over one of the central elements of modern science, namely the principled distinction between the objectivity of natural science, on the one hand, and human thought and action, on the other. To start with, everyone knows what it is like to see the yellowish colour of the leaves of trees in autumn. It is a sensory experience of the world. To have sensory experience is to take a perspective on the world that is necessarily a subjective perspective. A being that has sensory experience is not merely an object that moves according to natural laws, but it feels in a certain way to be in the world and to move in the world: seeing the colours of the leaves on the trees in autumn, hearing birds singing, tasting a ripe cheese, smelling roses, touching the surface of a wet stone etc. Natural science can reveal the sufficient physiological conditions for us to have sensory experience and the sufficient physical conditions for tree leaves to exhibit yellowish colours. It can even establish that there is no difference in the qualitative content of sense experiences without there also being a difference in brain states. Yet, it leaves the qualitative content of our sense experiences unexplained, that is, what it is like to have the experience of seeing the yellowish leaves of the trees in autumn (see Levine 1983). No functional definition can capture this experience. In philosophy of mind, this is known as the "hard problem of consciousness" (Chalmers 1996).

Subjective experience pertains to many higher-level living beings. As far as we know, thinking and acting are distinctive of us humans. Thinking and acting presuppose a subjective perspective on the world and thus sensory experience. Nevertheless, they are categorically different from sensory experience. Thinking and acting bring in *normativity*. It makes no sense to ask for a justification for the behaviour of animals. If a cat plays with a mouse before it catches and eats the mouse, one cannot condemn it. The cat behaves according to its instinct. It cannot be held accountable for its behaviour. In the case of a human being, by contrast, it makes sense to ask for reasons for the behaviour. This draws a line between scientific explanations and justifications (see Esfeld 2020, chapter 3, for details and references).

The problem is then that the extension of functionalism to human thought and action leads to a misapplication of science that trespasses its limits: one can provide a functionalist description of the manifested judgements and intentions of a person or group of persons at a time in terms of a role they played in their behaviour. But functionalism cannot say anything about whether these judgements and intentions to act were *correct*. In other words, it cannot say anything about what a person *should* think or do. It follows that modern science cannot say anything about this either. To return to the example of functionalism in ethics, the functional definition of the moral attitudes of people in a society at a time in terms of behavioural dispositions obviously does not and cannot answer the question of whether these attitudes are *right*. People within the society in question can also ask this question at any time. Functionalism therefore misses what characterises human thought and action, namely being accessible to reasons.

On the one hand, to a good extent, it is the success of modern natural science that leads to envisaging its unrestricted application, succumbing to the temptation of scientism. On the other hand, however, scientism is not the completion of science. Quite to the contrary, *scientism entails the self-destruction of science as a result of transgressing the borderline between objectivity and subjective experience and evaluations, thereby transgressing notably also the borderline between facts and norms*. Friedrich von Hayek (1952), in work going back to the exchanges with Karl Popper since the middle of the 1930s, but published only later, therefore calls scientism the "counter-revolution of science" and characterises it as "the abuse of reason", because the extension of natural science beyond its domain leads to polluting the scientific spirit with dogmatism and ends up in tyranny (Hayek 1952, p. 13). Karl Popper also published his philosophical criticism of scientism only after the publication of *The Open Society and Its Enemies* (1945) in a book entitled *The Poverty of Historicism* (1957).

Western civilisation, along its history, faced the branching of two paths at least twice. As we mentioned before, ancient Greece saw the branching of the Platonic and the Socratic-Aristotelian path. In modernity, Descartes' *Discourse on Method* (1637) marks a new crossroads. Not only modern science but also scientism branches off from it. When looked at it carefully, scientism is a reiteration of the Platonic path of comprehensive knowledge claims that reach all the way to grasping the idea of the universal good.

However, whereas such knowledge is conceivable in the Platonic framework of knowledge being the acquaintance with ideal forms, normative knowledge of the universal (or common) good is in principle not conceivable within the framework of modern natural science.

Objectivity is the hallmark of modern science: it abstracts from all subjective features. It attempts to take a standpoint from nowhere, that is, to reveal the facts as they exist independently of subjective judgements. Consequently, disciplined scepticism is the method of modern science. Yet, it follows from this alone that modern science meets a principled limit in everything subjective. If the scientific standpoint consists in refraining from all subjective elements, then the consequence is that one cannot grasp the subjective features with this standpoint. In other words, *modern science in the sense of natural science cannot, as a matter of principle, grasp its presuppositions in human thought and action.* Seeking to do so amounts to taking the other path, the one that departs from modern science. It leads to its self-destruction as we shall show in the following sections.

2.3 Scientism: A Brief Historical Background

René Descartes is a dualist. He conceives the human mind as a substance of its own, a *res cogitans* in contrast to the natural world as *res extensa*. Thomas Hobbes, his contemporary adversary in the mid-seventeenth century,[2] is a materialist. According to Hobbes, everything that exists is matter in motion. However, this ontological stance alone is not sufficient to turn Hobbes into an adherent of scientism. He does not seek to use modern science to derive norms and values that can be applied to society. In other words, Hobbes does not claim that modern science provides the answer to all meaningful questions. Together with John Locke, Thomas Hobbes is one of the founding fathers of the liberal tradition in modern social and political philosophy, as we shall explain in Chap. 4, Sect. 4.3 when going into the dilemma that Hobbes sets out for the modern constitutional state (and see Esfeld 1995 for details on the relationship between natural and political philosophy in Hobbes). Scientism comes up only later.

[2] The exchange between Hobbes and Descartes is documented as the *Third objections and replies to Descartes' Meditations*.

The term "scientism" itself with the meaning that has since become common appears to our knowledge for the first time in a statement of the French biologist Félix le Dantec published in 1911:

> I believe in the future of the Science: I believe that the Science, and the Science alone, will resolve all the questions that make sense; I believe that it will penetrate to the arcana of our sentimental lives and that it will even explain to me the origin and structure of the hereditary anti-scientific mysticism that coexists in me with the most absolute scientism. But I am also convinced that men ask themselves many questions that mean nothing. Science will show their absurdity by not answering these questions, which will prove that there are no answers to them. (Our translation of the French text as published on http://agora.qc.ca/Dossiers/Scientisme)

But the idea comes up earlier. At the turn of the eighteenth century, the effects of the French revolution on education as well as the rapid success of the natural sciences in technology, leading to what is known as the industrial revolution, give rise to the view that "The Science" may well pull the strings of society. A growing body of expert scientists and engineers starts to devise a prosperous society moulded according to "The Science". Henri de Saint Simon and Auguste Comte are paradigmatic figures of this extension of the natural sciences to society, absorbing moral sciences in the middle. In this spirit, Henri de Saint Simon writes:

> It is necessary that the physiologists chase from their company the philosophers, moralists and metaphysicians just as the astronomers have chased out the astrologers and the chemists have chased out the alchemists. (Saint Simon 1865–1878, vol. XV, p. 39; English translation quoted from Hayek 1952, p. 120)

In the search for the ideal of a "social physics", scientifically trained minds are in the business of extending their techniques and methodology not only to study society but also to organise it. Once again, Saint Simon is clear about this. In the face of the many questions and problems that societies are confronted with, he claims that "The Science" has the answers and the solutions:

> The decisions must be the result of scientific demonstrations totally independent of human will, and they will be subject to discussion by all those sufficiently educated to understand them (...) all social functions will be

> necessarily entrusted to those men who are most capable of exercising them in conformity with the general aims of the community. (Saint Simon 1865–1878, vol. XX, pp. 199–200; English translation quoted from Hayek 1952, p. 134)

This quotation is a paradigmatic expression of political scientism. As mentioned at the end of the previous section and as is evident from the quotation from le Dantec above, scientism is the claim that science can answer all questions such that it is supposed to set also the norms for human thought and action. When this happens in the political sphere, it is *political scientism* or social engineering. Like the art of technical engineering based on the discovery of the laws of motion of matter, there is supposed to be an art of social engineering that steers society through scientific knowledge.

However, there is a crucial difference: in technical engineering, the goal comes from outside science. People may, for instance, have the desire to travel quickly to other continents. Hence, they want engineers to build airplanes. To do so, the engineers have to know the facts about gravity. The possibility to construct airplanes is an example of how science enlarges the range of options that we have to self-determine our lives. But science does not prescribe what use we should make of these options. As another evident example, consider quantum physics. It discovers facts about radioactive atoms. This discovery enables us to make use of radioactivity for the purpose of energy production (nuclear energy) as well as for military purposes (nuclear weapons). But from quantum physics as a physical theory obviously follows nothing as to whether we *should* build nuclear power plants or produce nuclear weapons. Whereas in technical engineering the goal is imposed from outside science, social engineering pretends that science can prescribe goals not only for the lives of individuals but also for society as a whole: there is a common good for society, science tells us what the common good is, and it legitimises realising the common good through coercive political measures.

The most well-known historical case of social engineering, stemming from mainstream science and the middle of society, is the eugenics movement. While the idea of eugenics is old—it is, for instance, already present and implemented in Plato's ideal society steered by the philosopher-kings as set out in *The Republic* (see Sect. 2.1)—it was systematically implemented through social engineering in the first half of the twentieth century both in liberal democracies such as the United States and the

Scandinavian countries and in totalitarian regimes such as national socialism in Germany.

At the beginning of the twentieth century, the following was the majority view, if not quasi the consensus in biology, the social sciences and beyond: (i) There are superior and inferior hereditary traits. (ii) These are each distributed among certain social or ethnic groups. (iii) Cultural evolution in civilised societies undermines the mechanisms of natural, biological selection that hinder the spread of inferior hereditary traits. (iv) There is therefore a threat of the demise of civilised humanity unless the political authorities intervene to prevent the reproduction of inferior hereditary traits; in other words, the generation living in the 1920s is so to speak the "last generation" of civilised humanity, if not immediate action is taken to stop the people with inferior hereditary traits from reproducing themselves. (v) These interventions are necessary in virtue of the biological laws of nature concerning the selection of good hereditary traits and are therefore legitimised by these laws of nature.

Distinctive of this argumentation is the smooth transition from knowledge about biological facts—there are genetic differences that contribute to explaining behaviour, including cultural differences—to normative knowledge that legitimises political action that overrides basic human rights. That political action is moreover urgent and to be carried out without further debate, since we face a dire emergency, namely the demise of civilised humanity. The corresponding political action consisted primarily of sterilisations of those who were regarded as bearers of inferior hereditary traits. Legally, however, these were usually not explicit forced sterilisations. Rather, the consent of the victims was de facto enforced by threatening them with serious disadvantages if they refused, such as exclusion from social life, and the prospect of rewards if they "Follow the Science" and acquiesce in being sterilised: gaining freedom through sterilisation.

De facto enforced sterilisations were in particular practised in Western democracies that followed alleged science, mainly the US and the Scandinavian democracies. They were advocated across the political spectrum, including left-wing and liberal parties. By contrast, authoritarian-fascist regimes such as the one of Franco in Spain or the one of Salazar in Portugal took few if any eugenic measures because they paid no attention to alleged science as a guide in political decision-making.

In Latin America, the early twentieth century also saw various eugenics movements. With a strong dominance of Neo-Lamarckism, eugenics was regarded as a necessary means to control and to plan the immigratory European wave in the first two decades of the twentieth century. The idea of a "biological custom" aimed to preserve and to intensify the biological and cultural traits of "strong" races, avoiding hybridisms and harmful crossbreeding. Many scientific associations (such as the "Sociedad Argentina de Eugenesia", founded in 1912) and governmental policies typified some cultural, ethnical and biological traits as "dysgenics" (Delfino 1926; Verano 1929). While eugenics in Latin America initially aimed to control and plan immigratory waves according to the idea of a "healthy civilised nation", adopted mostly by young liberal governments and their aristocratic elites, by 1930 Latin American eugenics changed to a "Pan American ideal", to resist the influence of the United States in the region. Surprisingly, the ideal of a "strong race" changed from Anglo-Saxon to Italian traits, more long-lived and fertile than others (Maiocchi 1999), in the frame of closer relationships between Latin American governments and the Italian fascism (Stepan 1992).

Germany under national socialism, in turn, drove eugenics to the consequence of exterminating entire ethnic groups with allegedly inferior hereditary traits in order to realise the goal of a racially pure society—a goal which the national socialist ideology took to be predetermined by science, namely by the laws of biology. The scientists who advocated eugenic measures in the West had of course nothing of that kind in mind. However, the lesson to be learnt from the history of eugenics is this one: once one places science above human rights by taking the scope of science to be unlimited and by considering it to be in the position to legitimise political actions that override basic human rights of defence against interventions into one's course of life, including one's bodily integrity, there are no boundaries anymore. If science claims to take precedence over basic human rights, there's nothing left that can stop the justification of any atrocities under the pretext that they are imposed by science. This is the ultimate consequence of "Follow the Science" that the history of eugenics in the twentieth century illustrates: *"Follow the Science" amounts to using science as a weapon against human rights.*

The "Follow the Science" slogan encloses quite well the essence of scientism. In a nutshell, it refers to the attitude to regard the methodology and techniques of the natural sciences not only as the standard of knowledge but also as the guideline for actions, applicable to all domains without limits. It thereby trespasses the frontier between facts and norms. In doing so, scientism professes a dogmatic attitude towards scientific knowledge in the form of unquestionable scientific consensuses. It enshrines a group of experts as the representative of "The Science" who determines which views are true and which ones are false. It epistemically and morally judges citizens' attitudes and behaviours depending on their compliance with what "The Science" says and commands.

As we mentioned in passing before, scientism rarely comes on its own; rather than being an isolated phenomenon, it is in general used by political power to justify coercion. This, then, is the distinctive trait that transforms scientism into political scientism. That notwithstanding, it is worth drawing the conceptual difference between the foundations of scientism as an attitude towards the natural sciences and its scope, and political scientism as its implementation to justify coercion.

As the case of national socialism in Germany shows, there is a smooth transition from "Follow the Science" into adopting coercive political measures that override basic human rights to prevent an allegedly apocalyptic future such as the demise of civilised humanity to an outright totalitarian regime. Indeed, the totalitarian socialist regimes of the twentieth century both of the national socialist and the communist type are cases of political scientism. National socialism considered itself to be strictly scientific, implementing only the biological laws about the evolution of races that point to the pure-blooded society as the ultimate goal of history. By the same token, communism relied on alleged scientific laws of the evolution of matter that apply also to human societies and that inevitably lead to the classless society as the ultimate goal of world history. Both pursued the programme of transforming the existing human beings into allegedly morally better humans that meet the standard of the final goal of history as established by science in the guise of the pure-blooded or the classless society. Both recognised no limits to scientific knowledge.

Indeed, political scientism, social engineering and the slogan "Follow the Science" are far from being mere technocratic solutions to practical problems that arise in politics. They are an encompassing programme to turn the existing human beings into allegedly morally better humans by implementing a common good that science is supposed to discover, whatever the science in question and the common good at stake then may be in concrete terms. This makes political scientism so dangerous. The lesson of the history of the twentieth century from the eugenics movement to the totalitarian socialist regimes of the national socialist and the communist type is clear: *once claims to scientific knowledge are employed to derive political demands that override human rights, there is no longer any normative limit*. The door is then opened to do anything in the name of science, from plunder and planification to the outright murder of human beings.

What is hence surprising is not that we find traces of analogues of the "Follow the Science" slogan back in the early twentieth century, when the success of modern science in discovering the laws of the motion of matter led to an enormous technological progress with the industrial revolution. What is rather surprising is that we today, in the twenty-first century, still revisit these old places and are influenced by them. Political scientism is back in the middle of our societies with the corona and the climate regimes. These regimes, in turn, are fed by intellectual postmodernism and its offspring, the wokeness regime as we shall explain later. In short, the "Follow the Science" slogan is not a problem because it is new and unknown. It is a problem because it is old and well known: we know what its disastrous consequences were, and we see these consequences again unfolding today.

To start with, let us consider three examples that are symptomatic of the return of political scientism. On 16 August 2020, a woman was violently arrested by five officials for taking her dog for a walk on Sunday noon, in the small city of Bariloche (Argentina), in the sparsely populated Patagonia.[3] The reason for the detention was that the woman violated the "Aislamiento Social Preventivo y Obligatorio" (Mandatory and Preventive Social Isolation), which banned people from venturing outside on Sundays. The justification for that coercive measure was that "it was necessary to protect population's health from Covid-19 infection" as reported by the Bariloche city major. It might be thought that this is an outlandish case, but it comes from one of the most strongly democratic countries in Latin America. Unfortunately, cases like this were very frequent under the pretext of protecting people's health based on alleged scientific knowledge about a new type of coronavirus. For instance, the Peruvian government issued a State of National Emergency on 15 March 2020 as a response to an increasing COVID-19 infection rate in Europe. The national lockdown forbade citizens to venture outside their homes at any time unless it was strictly necessary and duly justified. On 20 March 2020, a Peruvian man in pyjamas was arrested in front of his home for taking his dog for a

[3] https://www.lanacion.com.ar/sociedad/coronavirus-argentina-mujer-salio-pasear-su-perro-nid2423060/

three-minute walk. To add insult to injury, not only was he arrested but the dog was also confiscated.[4]

In February 2023, the European Parliament voted to pass a law banning the sale of new diesel and petrol cars from 2035 on. The idea, as the President of the Transport Committee of the European Parliament, Karima Delli, reported, is to "reconcile the automobile and the climate—two rival brothers, so to speak".[5] The car ban law is part of the European strategy to combat climate change, again based on alleged scientific knowledge about human-made climate change leading to an apocalyptic future unless governments intervene with draconian coercive measures. Thus, car buyers will no longer be allowed to purchase gasoline-fuelled cars, and car manufactures will no longer be allowed to sell them. Putting aside concerns about freedom to buy and sell, the law puts the entire European car industry at risk, affecting mainly the lower and middle classes. The car industry employs 270,000 workers only in Italy. Paradoxically, luxury cars (such as Ferraris) will be exempted from the law, clearly benefiting the better off. However, once again, this is not an exemption: many countries have implemented legislation to fight climate change, controlling not only *what* may (and may not) be produced, bought and sold but also *how* it must be produced, bought, and sold.

In 1950, the University of California at Berkeley imposed on its employees a loyalty oath towards the constitution of the United States and of the State of California, which included a declaration of no affiliation to "the Communist party or any other organisation which advocates the overthrow of the Government by force or violence".[6] History correctly remembers this period as the paranoid McCarthy-era persecution of alleged communist sympathisers in the context of the Cold War. After more than 70 years, UC Berkeley imposes on its employees now a self-authored "Diversity and Inclusion Vow". They must declare knowledge and experience with dimensions of diversity that result from different identities; such knowledge must also be non-vague, inclusive and willingly informed (see Caplan 2021). Will history remember these days as the paranoid wokeness-era persecutions and cancellations? University vows are only a tiny part of

[4] https://www.clarin.com/viste/coronavirus-peru-paseo-pijama-perro-detuvieron-familia-grito-llevaran-mascota_0_u9P8Vg-Pg.html

[5] https://www.brusselstimes.com/371290/eu-officially-bans-new-diesel-and-petrol-cars-from-2035

[6] https://oac.cdlib.org/view?docId=hb0x0nb40m&brand=oac4&doc.view=entire_text

a growing cultural wave that has swept Western societies over the past two decades.

A confusion between the private and the public spheres has put individuals in an always-under-suspicion situation of openly and explicitly declaring their beliefs about race, gender, social justice, the environment, identifiable groups' rights, history etc., as a way to show that they are on the right side of morality and history, as a contemporary proof of virtue. Failure to recognise even the importance of these principles may mean cancellation, public punishment or social ostracism. Thomas Sowell, in his 1995 book, *The Vision of the Anointed*, warned of a growing intolerant and self-indulgent view among academics and intellectuals. In less than three decades, a new value system made its way from the Ivory Tower into society, creeping into political power and the legal system (e.g. see Canadian Bill C-16 in force since 2017).[7] Mandatory educational courses on gender and inclusion for government employees (such as the "Ley Micaela" in Argentina in force since 2019),[8] or an extended witch-hunting policy in the name of anti-racism, inclusion, diversity and equality are some of the instruments that tribalise societies, silence diversity, invade the private sphere and reaffirm the extension of government policies to combat non-compliance with these new values.

These three concrete cases would merely be unfortunate anecdotes if they were not integral parts of bigger narratives. Rather than exemptions, they have become common currency in our lives. In the form of mandatory isolations, nation-wide lockdowns, vaccination passports, carbon print trackers, or a fierce opposition to dissensus, our lives are increasingly affected by, on the one hand, a steady abandonment of the method of science, and, on the other hand, the demise of the rule of law and the respect of human freedom and dignity. Instead, authoritarian reactions to a wide panoply of "dire" emergencies that undermine peaceful cooperation and free judgement along with the marketing of postfactual realities that erode scientific objectivity and hamper scientific discussion have largely become the heart of small narratives that pervade societies at large. In short, we undergo the return of scientism and its political use, ushering in a trend towards a new, specifically postmodern totalitarianism that we dub "actually existing postmodernism". We will substantiate this claim in the following four sections by going into the corona regime (Sect. 2.4), the climate

[7] https://www.parl.ca/LegisInfo/en/bill/42-1/c-16?view=about
[8] https://www.argentina.gob.ar/generos/ley-micaela

regime (Sect. 2.5) and the wokeness regime (Sect. 2.6). Finally, we will pull all these strings together in showing how these regimes establish a new version of political scientism and its totalitarian consequences (Sect. 2.7).

2.4 The Corona Regime: The Postfactual Reaction to the Spread of a Virus

The corona regime—the political reaction to a virus wave with comprehensive regulations of social and even private life—exhibits all the characteristic features of political scientism, as we shall show in this section. This book, however, is not about the scientific and political reactions to the coronavirus waves in particular. We use this case only to illustrate scientism and its political use.

To begin with, let us recall some pieces of what was established in science and medical practice *before* 2020. In 2019, in the frame of the Global Influenza Programme, the World Health Organization (WHO) published a thorough document providing recommendations for the use of non-pharmaceutical public health measures for mitigating risk and impact of epidemic and pandemic influenza (World Health Organization 2019). The recommendations, which are addressed to national governments and centralised agencies for public health, are based on an exhaustive review of the existing scientific literature until 2019. The document, published just a couple of months before the outbreak of a new coronavirus was declared in China, contains some interesting results. For instance, it advises against any contact tracing policy. In part, contact tracing is pointless for airborne viruses since the velocity of propagation is too high; but it was also not recommended because the quality of the scientific research was very low. Even more striking is that the document does *not* recommend the quarantine of exposed individuals. It does not advise border closures either, which may cause the immobilisation of resources needed in an emergency. Other non-pharmaceutical interventions are only conditionally recommended, circumscribed to early stages of a virus outbreak (e.g. internal travel restrictions). In most cases, the document states that the quality of the evidence is low or very low.

Despite these recommendations, the Chinese central government imposed a city-wide lockdown on Wuhan on 23 January 2020. The measure was absolutely unprecedented and contrary to all science-based recommendations. Gauden Galea, WHO representative in China, said on the very day that the Wuhan lockdown was imposed: "The lockdown of 11 million people is unprecedented in public health history, so it is certainly not a recommendation that the WHO has made".[9] Galea was just echoing from the 2019 WHO document. Lawrence Gostin, public health professor at Georgetown University in Washington, DC, declared in the same vein: "The truth is that these kinds of lockdowns are very rare and never effective".[10]

Against this background, when from February 2020 on, politicians also in Europe contemplated the idea of locking entire cities down (e.g. see the statement of the British health secretary Matt Hancock on 1 March 2020),[11] our initial reaction was that scientists would speak out and that the media and the public would oust politicians who tried to copy the Chinese policies in the Western world. We were, of course, completely wrong about that. Soon, we were afraid not of a virus, but afraid of the reaction to the spread of the virus in science and politics.

Despite science and the rule of law, not only established evidence and what was hitherto the state of the art in medical science were ignored, but also, like in an Orwellian novel, scientific standards were abandoned and the very meaning of the terms was distorted. Consider the very concept of a pandemic. The term "pandemic" suggests the spread of an extraordinary virus wave associated with an extraordinary danger to the general population. However, already in May 2009 in the context of the swine flu, the WHO stated that to declare a viral wave a pandemic, no excess mortality affecting the general population is required. Infections with a virus that is spreading worldwide and that the population is not considered to be immune to are sufficient (see Doshi 2010; Keil et al. 2011). The eminent German epidemiologist and then WHO director Ulrich Keil pointed out at a hearing of the Council of Europe in 2010 that in this way annual flu waves can be declared a pandemic based on the number of infections alone (Keil 2010, p. 2). With this use of the term "pandemic", it is therefore possible to stage a virus outbreak that is not dangerous for the general population as a pandemic at will, thereby stirring up fear and panic in the population. Infections are sufficient.

[9] https://www.reuters.com/article/us-china-health-who-idUSKBN1ZM1G9

[10] https://www.washingtonpost.com/graphics/2020/world/corona-simulator/

[11] https://www.theguardian.com/world/2020/mar/01/matt-hancock-ministers-to-publish-coronavirus-battle-plan-for-uk

As far as the infections during the COVID-19 pandemic are concerned, in order to detect them, medical standards in diagnostics were distorted. Unlike all other diseases, no differential diagnosis took place in the case of a suspected infection with the new coronavirus. A positive test was sufficient, namely the positive result of a polymerase chain reaction test (PCR). Indeed, the PCR test rapidly became the gold standard to identify infections in a population. However, the inventor of this test, Nobel Prize winner Kary Mullis, had warned against using the PCR test as a benchmark for diagnosis. At a conference, he said in an answer to a question from the audience that "with the PCR, if you do it well, you can find almost anything in anybody" (Mullis et al. 1997). The reason is that the PCR test is not able to pick out a specific disease pattern. This does not prevent it from being used for diagnostic purposes, but only when accompanied with the screening of patients' symptoms.

This change in the medical criteria for diagnosis distorts the data severely. The detailed examination of the results of 162,457 PCR tests evaluated in a laboratory in Münster in Westphalia (Germany) revealed that only 40.6% of those who tested positive had a viral load sufficient to contribute to the spread of the virus (Stang et al. 2021). Nevertheless, all individuals who tested positive were deprived of their liberty and quarantined based on the PCR test alone, without any examination of whether they could transmit pathogens to others or had any symptoms of a disease at all. Thus, even persons without any symptoms of the disease were declared to be sick with COVID-19 based solely on a positive PCR test and were subsequently certified to have recovered from the disease, even if they had always been in perfect health. Similarly, all persons hospitalised and all deceased persons with a positive PCR test were counted as COVID-19 hospitalisations and COVID-19 deaths—without any investigation of whether these persons were sick with COVID-19 at all or whether other ailments led to hospitalisation and, if so, death.

It also fits this framing that there is no specific clinical picture for the infection with the supposedly new coronavirus. The COVID-19 disease, caused by the SARS-Cov-2 virus, is a respiratory disease with symptoms typical of this class of disease. COVID-19 is not a new type of disease for which there is a specific medical classification—except just the PCR test (see Afriat et al. 2023). COVID-19 became a pandemic only because of the PCR test and not because of epidemiological phenomena specific to the virus.

In short, there is the virus SARS-Cov-2 and its specific microbiological classification based on certain molecular characteristics. There are the infections, hospitalisations and deaths caused by the virus. However, something new grew out of this only by the reinterpretation of the central terms like "pandemic" and "infection". In that sense, the coronavirus pandemic is a *postfactual* pandemic, based on the distortion of language and, as we shall explain, the replacement of reality as given through empirical data with model simulations.

The paradigm shift from a medicine-based to a politics-based strategy to deal with the spread of a virus happened between February and early March 2020. In particular, the paper by Neil Ferguson and collaborators from Imperial College London, published on 16 March 2020, set a breaking point in this story. This paper had a significant impact on the actions of the UK and US governments, and probably a number of other Western governments as well. Two aspects can be observed. The first concerns the quality of the study as a *scientific* paper. We discuss the key points in that respect in what follows. More importantly, the paper is a clear case of scientism and political scientism. Our main focus is on that aspect. Since this is not a book on COVID-19 in particular, we cannot give here an overview of the vast literature on that subject. We pick out the study by Ferguson et al. (2020) in order to show how the reaction to the coronavirus waves from spring 2020 on brought scientism and its political use back on the agenda. That is, our objective in picking out this study only is to illustrate scientism and its link to politics.

The predictions of the paper were indeed bleak. It forecasted that without non-pharmaceutical interventions (NPIs), the United Kingdom will reach 510,000 excess deaths from COVID-19 and the United States 2.2 million by the summer of 2020—that is, the shocking amount of more than 4000 and 16,000 excess deaths per day, respectively. To prevent such a dreadful scenario, coercive NPIs were required: the primary goal of such measures was to "flatten the curve", slowing down the spread of the virus to avoid the collapse of public health systems and massive deaths due to the unavailability of proper health care. A series of intermittent lockdowns was recommended until a vaccine was available and administered to most of the world's population. Ferguson and collaborators estimated that this would take approximately 18 months from March 2020 on. In a nutshell, the paper projected an exorbitant number of deaths in a few months, made the case that draconian, coercive measures had to be implemented to mitigate the effects of the outbreak and conditionalised the beginning of the end of the pandemic (and the draconian NPIs) on the success of a central planning strategy: development of a vaccine and inoculation campaigns.

In the following, we focus on the six most important flaws in Ferguson et al.'s paper that illustrate how this paper implements the transition from science to scientism as explained in Sect. 2.2, namely the abuse of models and the misapplication of functionalism to human action, thus paving the way to political scientism.

(1) *Poor and overestimated initial data of fatality rate*: To begin with, Ferguson et al. assumed that COVID-19 was dangerous for the *entire* population. This assumption was unfounded. From the beginning, the available data showed that the infection fatality rate (IFR) increased significantly with age, putting at risk *only* the population over roughly 70 years old and with a weak immune system. In fact, the average age of people who died and had a positive PCR test is well over 80 years in most Western countries, that is, higher than the average life expectancy rate. In a more complete study carried out later, it is concluded that for the age groups below 70 years, the overall IFR was between 0.05% and 0.1%; that is, five to ten deaths per 10,000 infections (Pezzullo et al. 2023). This corresponds to general life risks such as riding motorbikes, regularly driving longer distances by car, doing sports like hiking, skiing, climbing etc. No one considers banning these activities on the basis of such a tiny risk.

The initial parameters that are introduced in the models to project possible outcomes deserve special attention. In the modelling of the evolution of an epidemic, a reliable IFR is crucial: it basically tells us how many people are expected to die after being infected. Ferguson et al. started off with an extremely high IFR, around 0.9%. Yet, as early as March 2020, John Ioannidis (2020a) estimated on the basis of the data available at the time—for instance, the systematic tests of all passengers on the Diamond Princess cruise ship—an IFR for the entire population that was clearly well below 1%. In Ioannidis (2020b), the IFR of more than 50 locations is analysed, setting the corrected median IFR at 0.27% (see also Axfors and Ioannidis 2021). In Germany, Hendrik Streeck and colleagues tested the entire population in the village of Heinsberg in April 2020 and came up with an IFR of 0.36% (Streeck et al. 2020).

Hence, the initial projections of Ferguson et al. were unfounded because the data for the IFR in their model was overly inflated to begin with; this was already evident at the time of the publication of the paper, even if the data were initially quite poor and only partial evidence available. However, this implies in any case that it is irresponsible—if not morally outrageous—to use poor evidence as the basis for a policy advice in the form of coercive measures that severely violate basic human rights. On the one hand, it was well known from the outset that implementing such measures will cause a lot of harm, including deaths. On the other hand, there was no evidence for such measures being appropriate or having any significant effect at all, given the above-mentioned state of the art in medical science before 2020. Thus, Ron Paul aptly wrote on the very day when the paper by Ferguson et al. was published, 16 March 2020:

> If anything, what people like Fauci and the other fearmongers are demanding will likely make the disease worse. The martial law they dream about will leave people hunkered down inside their homes instead of going outdoors or to the beach where the sunshine and fresh air would help boost immunity. The panic produced by these fearmongers is likely helping spread the disease, as massive crowds rush into Walmart and Costco for that last roll of toilet paper. … People should ask themselves whether this coronavirus "pandemic" could be a big hoax, with the actual danger of the disease massively exaggerated by those who seek to profit—financially or politically—from the ensuing panic.[12]

This assessment proved to be entirely correct later. But it was already clear from the beginning.

(2) *Underestimation of the limits of modelling*: Ferguson et al. advocated their computer modelling as a reliable representation of reality. Yet, as explained in Sect. 2.2, scientific models—and computational models in particular—are not representations of reality. Their predictions crucially depend on which parameters are taken into account, how their initial values are set and how many intervening elements are idealised. There always is some leeway to set the value of the initial parameters. Slight differences in the initial values can lead to huge differences in future scenarios that the

[12] http://ronpaulinstitute.org/archives/featured-articles/2020/march/16/the-coronavirus-hoax/

models predict. It is a mistake to confuse them with knowledge of reality and of what will happen in the future.

We are all familiar with these shortcomings of models when it comes to forecasting the weather or predicting the future price of assets in a market. Banks regularly publish forecasts at the beginning of a year about what the value of financial instruments such as shares, bonds, commodities, precious metals, currency pairs etc. will be at the end of the year. However, such forecasts regularly turn out to be only an estimation under stable conditions, based on a selected group of relevant parameters and a huge amount of data gathered for years. Any investor knows that these projections do not assure knowledge. For instance, it is impossible to know all the future-intervening parameters in a market that involves millions of persons. In short, no computer-modelling projection can claim to accurately represent reality, let alone reliably predict what will happen in the future. It might deliver psychological comfort to the wish for knowledge and control of the future, but this is ultimately an illusion.

Modelling a virus epidemic is not any different. In setting the initial values of crucial parameters such as the IFR, the number of symptomatic cases requiring hospitalisation or treatment in intensive care units very high, Ferguson et al. neglected the range of possible values for these parameters in an initial stage of uncertainty. This shortcoming casts doubts on their purposes when releasing the paper: if they had said that the parameters allow for a range of possible initial values and that consequently there is a wide range of possible outcomes for the evolution of the virus outbreak, then they could not have provided bleak projections that could serve as a basis for the justification of coercive political measures such as lockdowns.

(3) *Omission of people's spontaneous adaptation to new information*: People react sensibly to changes and challenges. They adapt to new situations and modify their behaviour in the light of new information. Ferguson et al. simply omitted this fact in their model, applying in fact a crude version of functionalism to human action. People were modelled as mechanical bodies, bouncing blindly around and infecting each other. This emulates models in physics. For instance, physicists approximate the behaviour of real gases in terms of the behaviour of ideal gases. This allows them to identify qualitatively similar behaviours in a more tractable theoretical construct. In an

ideal gas model, the molecules of the gas are point-like rigid spheres, perfectly elastic and do not interact with each other (they only elastically collide with each other and with the container's walls).

While this assumption is useful and harmless in physics, it is highly misleading when modelling people's behaviour. People's adaptation to new information is then not a trivial omission. Quite to the contrary, this is the parameter that is most important when transferring models from the natural sciences to the humanities and social sciences: people are not physical objects whose trajectories can be modelled by merely taking external factors into account. If a new virus hits a population, people adapt their behaviour on their own, taking care of themselves and their families, friends and neighbours. When it is projected that infections will increase exponentially indefinitely, which is at the basis of the prediction of massive deaths and collapses of public health systems, we know from the outset that such a claim is implausible and that the model thus is severely flawed. People think and act. When they receive information about a virus outbreak, they change their behaviour on their own. Accordingly, the exponential increase in infections had already been broken when the lockdowns were imposed (e.g. see Kuhbandner et al. 2022). Again, this was known all along: there was never an unlimited exponential increase in infections during any epidemic in history.

Yet, how people adapt their behaviour to new information cannot be predicted. That is why functionalism applied to human action fails when employed as a predictive tool. It is not possible to know in advance what strategies people will invent and adopt to deal with a new situation. It follows from this that any model that predicts a static or linearly progressive future—and that includes a linear exponential growth of infections—overlooks the creativity of human beings in developing new strategies to deal with new scenarios. Consequently, the simple application of methods from the natural sciences—such as predictions based on models—fails in the humanities and social sciences. Friedrich von Hayek (1952) and Karl Popper (1957) drove this point home more than 60 years ago, as mentioned in Sect. 2.2. Awareness of Hayek's and Popper's works on the methodology of science alone would have been sufficient to prevent much of the cartoonish, simplistic science that has been presented to us as "The Science" in the past years.

(4) *Failure of scientific reasoning in the assessment of non-pharmaceutical interventions' consequences*: As mentioned above, if there was a "scientific consensus" back in 2019, it was one that discouraged coercive, long-lived non-medical interventions on populations. It was also one that explicitly stated that the scientific evidence in their favour was absent. A targeted and medical strategy was thus to be preferred as set out in the *Great Barrington Declaration* released on 4 October 2020: focused protection. As the initiators of that declaration—Jay Bhattacharya et al. (2020)—repeatedly argued, a focused-protection approach aims to offer targeted protection to safeguard those at serious risk, without causing irreparable damage to the rest of the population.

Of course, no one possesses confirmed a priori knowledge of what the best strategy is to respond to a new situation. But then laissez-faire is the best strategy: when different strategies are tried out spontaneously, one will see which is the most appropriate way to deal with a new situation. This is not the endorsement of any "numbers trump ability" theorem. The point is that in a context of uncertainty and subjective risk assessment, trying out different strategies works better than centrally imposing one measure for all. Isolating oneself at home may be affordable in high-GDP-per-capita countries, like Switzerland; but it is plainly unbearable in a Brazilian favela (see Green 2021). It may also be more suitable for certain individuals, such as professors or state employees, as their salaries are monthly guaranteed by state resources; but it is unbearable for blue-collar workers whose income mostly depends on demand and economic activity. Generally speaking, if different strategies are tried out spontaneously from below and based on voluntary adaptations and interactions, it is also much easier to correct mistakes than when a centrally imposed strategy from above is implemented. As a rule, the latter is only abandoned when great damage has already been done. With a variety of strategies in the run, each strategy has to prove itself in the confrontation with other strategies and shortcomings can be quickly corrected.

It could be objected that such an approach is morally irresponsible and that governments should act on the basis of a principle of precaution. This reasoning is, however, flawed in many aspects. It, first, assumes that governments are morally responsible for people's welfare, thus presupposing a paternalistic conception of government. This in turn presupposes that individuals are unable to deal by themselves with the situation. It, second, suggests that the alternative to government action is do nothing, a "let it rip" approach. But this is fallacious. The genuine opposition is the one of centralised versus decentralised planning. It, third, opens the door to *any* action on governments' behalf: there always is some principle of precaution that can justify any government actions regardless of their costs and consequences.

To sum up, the following three elements have to be born in mind:

- Empirically, in the case at hand, as mentioned above, the available data established from the beginning that the statistical risk of serious health consequences from an infection with the coronavirus depended on the age, putting at risk only the elderly population with a weak immune system. There was no evidence that suggested a serious health risk for the general population. There never was any uncertainty about this matter.
- Methodologically, it is well established in the literature since Mises (1949), Hayek (1952) and Popper (1957) among others that spontaneous adaptation and decentralised planning trump central planning in coping with a new challenge. To put it in a nutshell, the local knowledge of the many supersedes knowledge that is centralised in the hands of a few. The reason is that a central authority can simply not accumulate all the knowledge about the different local situations and respond adequately to their variety.
- Normatively, it is a priori clear that government interventions through coercive political measures create harm for certain segments of the population. In the case at hand, it was from the outset evident that lockdowns have cruel, morally reprehensible consequences, such as forcing people in hospitals and homes for the elderly to die in isolation without being accompanied by their families, and that they lead to otherwise avoidable deaths (from psychological stress, failure of appropriate medical treatment of diseases etc.). In short, it was certain from the outset that lockdowns cause great harm, whereas it was uncertain what, if any, their benefits would be—given in particular that the state of the art before 2020 advised against lockdowns and coercive political measures for the population in general, as mentioned above.

Although there were attempts to enforce the lockdown strategy worldwide, the political reactions to the coronavirus waves were diverse. In

Western democracies, only a few US states such as South Dakota refrained entirely from imposing state-mandated coercive measures. This was also the reaction of some authoritarian states, such as Belarus, that is, states whose governments are not inclined to listen to—alleged—science. But Western democracies introduced coercive measures of widely varying severity and ended them at different times: measures ranged from relatively minor restrictions such as bans of mass gatherings to lockdowns involving business and school closures and prohibitions on leaving one's home, meeting friends and family, then later *de facto* or even legally binding vaccination mandates etc. Some states ended these measures after the first coronavirus wave in summer 2020, while others held on to them much longer.

This diversity allows us to assess the success or failure of the shift from the medical to a coercive political strategy. If there had not been a variety of reactions, then it would be impossible from the outset to prove the effectiveness of the political strategy. Consider the following analogy as a thought experiment: suppose that tribes in a possible world respond to drought with rain dances. Let us assume that this is a collective, political reaction to droughts: everyone is forced to join in the dance. Those who do not join in are seen as selfish and excluded from social life. If it does not rain despite dancing, then the dancing becomes even more intense. More and more intense dancing continues until rain eventually comes. If all the tribes do the same thing, namely dancing until it rains, then the tribal leaders can always praise the responsible behaviour of their fellow tribesmen and claim the following: if everyone had not danced so intensely, it would not have rained. Obviously, such a claim is worthless. If everyone does the same, it is impossible to obtain evidence for this claim.

That is why trying out different strategies has such a great insight value: it is the only way to find out which strategy works and which does not. That is why scientific studies always require a control group that is not exposed to the measures or interventions whose effectiveness is to be studied. If possible, the allocation to the two groups—who will be exposed to the measures or interventions and who will be in the control group—should be randomised. The researchers conducting the study should not know who is in which group.

When testing the effectiveness of medicines, including vaccines, this is possible. This is obviously not possible for measures such as wearing masks, restricting access to restaurants and the like, lockdowns with school and business closures etc. The control group then consists of some states or communities doing this and others not. Consequently, scientists and policymakers should welcome the fact that different states have responded in different ways to the coronavirus waves rather than calling for the global enforcement of coercive measures. After all, allowing different strategies is the only way by which the proponents of measures could prove their claims about their effectiveness. Scientists who oppose such control groups abandon scientific reasoning and revert to the level of tribal leaders who take the temporal succession of increasingly intense dancing and rain as evidence that the intense dancing *causes* the rain. Scientism and political scientism contradict the very principles of the scientific methodology by seeking to enforce one strategy worldwide.

If the political coercive measures have achieved their goal of protecting the population, then the following correlation should be demonstrable: states that have only taken a few measures, such as only bans on mass gatherings, should have seen relatively more hospitalisations and deaths associated with COVID-19. In brief, the more stringent the coercive measures, the fewer deaths and severe cases of COVID-19 should occur in relation to the size of the population. Of course, a simple linear relationship cannot be expected because many factors intervene. But if no pattern whatsoever emerges in the form of an overall correlation between the stringency of the interventions and the reduction of severe courses of disease and deaths, then we have no evidence that the interventions were effective.

This is indeed the result that has been established by numerous, independently produced scientific studies and a number of meta-studies that summarise the results of the individual studies.[13] If one considers Western countries in the Northern Hemisphere with comparable levels of economic development and health systems and assesses the relevant data on infection incidence such as hospital admissions and deaths relative to population size, it is not possible to infer from these data which countries have adopted severe coercive measures such as lockdowns and which ones have not. That is, countries with strict measures such as the United Kingdom, France, Italy and Spain did not overall perform any better than countries such as Sweden, which has taken only very few measures, and Switzerland, which has imposed considerably milder measures than its neighbouring countries. Although it is possible to find dissimilarities in, for instance,

[13] See in particular Bendavid et al. (2021), Bjørnskov (2021), Chin et al. (2021), Kowall et al. (2021), Herby et al. (2022, 2023), Mader and Rüttenauer (2022). See furthermore Bagus et al. (2022) and the overall assessment by Templeton (2023).

excess mortality rate locally and within a limited span of time between different strategies, when longer periods are taken into account, these dissimilarities vanish so that no overall pattern of a correlation emerges (see Forthun et al. 2024 for a comparative study of the Scandinavian countries). An analogous result can be found by comparing the performance of different states in the United States, as documented in detail in a report by the US *National Bureau of Economic Research* from April 2022 (Kerpen et al. 2022).[14] Most recently, a study by Eran Bendavid and Chirag J. Patel (2024) confirms on a very broad data basis that there is no overall pattern of a correlation between political measures to fight the coronavirus waves and positive health effects.

The conclusion hence is that the shift from the medical strategy of focused protection to the political strategy of coercive measures has failed. There is no proof of any significant effect of the adopted non-pharmaceutical interventions, whatever their severity was. This result also makes a cost-benefit analysis of the lockdowns and other coercive policy measures superfluous. These measures have not produced any benefit that can be proven to be significant in any way. They have only caused huge damages including deaths.

(5) *Reliance upon one particular solution from the outset instead of open-minded research*: As a treatment for COVID-19 to end the declared pandemic, the paper by Ferguson et al. relies from the outset on vaccination only. This, again, is a striking lack of scientific reasoning: one has to be open-minded and allow for different solutions to be developed and to compete with one another in order to find out which solution works best. The paper by Ferguson et al. manifests again Soviet-style central planning: vaccines were to be developed and administered to the population, for which Ferguson et al. estimated about 18 months from March 2020. This, indeed, was the only prediction in the paper that turned out to be roughly correct—but only as the implementation of a plan and not as something that was the result of research conducted according to scientific standards.

[14] See also the early studies by Raffelhüschen (2020) for Germany and Miles et al. (2020) for the United Kingdom. For Switzerland, see in particular Beck and Widmer (2020), Gerber (2022) and the contributions in Beck et al. (2022).

The only data on vaccination that was collected in line with scientific standards comes from the licensing studies for the vaccines. Scientific standards here means that there is a control group that receives a placebo instead of the vaccine. The study participants are divided into two groups according to the random principle (randomisation). The participants and the observers do not know who belongs to which group (blinding). However, instead of letting the trials run for at least one year to obtain meaningful data, the blinding was lifted after only two months of average observation time and the participants in the placebo groups were offered the vaccination. To cite ethical reasons for this turns ethics on its head: to claim that it is unethical to withhold vaccination from people in the control groups assumes that the efficacy and the safety of the vaccines are known. However, this knowledge can only be gained through the studies. It is therefore unethical to prevent us from gaining this knowledge by prematurely stopping the trials with volunteers who have signed up for them. In consequence, we do not have control groups that have been systematically observed over a longer period of time, and we can no longer obtain them.

An unbiased assessment of self-protection, external protection (no passing on of the virus) and safety of the vaccines is therefore only possible on the basis of the data from the average two-month observation period in the registration studies. The independent analysis of these data by Joseph Fraiman, Peter Doshi and others, published in August 2022, shows worrisome results: in terms of serious adverse events that occurred in the Pfizer/BioNTech and Moderna pivotal trials, there was an absolute risk increase from vaccination of 10.1 and 15.1 per 10,000 vaccinated participants, respectively, compared to the placebo group. A prevention of a severe course with hospitalisation was proven in 2.3 and 6.4 per 10,000 vaccinated participants compared to the placebo group. The analysis of the registration studies thus shows that the harm of vaccination outweighs the benefits when vaccination is applied to the general population, that is, independent of the personal risk of severe COVID-19 disease. In other words, more severe adverse effects of the vaccines occur than severe courses of COVID-19 are prevented by them. This negative benefit-harm ratio calculated from the data of the licensing studies available so far requires further clarification, in particular the anonymised release of the health data of the individual study participants (see also Pfizer Documents Investigations Team 2023). That notwithstanding, the assessment of Christof Prugger and co-authors, who summarised the situation with regard to vaccines in an article in the *British Medical Journal* at the end of 2021, is well-taken:

> Covid-19 vaccines were widely administered following "conditional" authorisation based on short clinical trials, when important questions remained unanswered. (Prugger et al. 2021, p. 3)

These important questions include, above all, the question of whether the vaccination is sterile—that is, whether the further spread of the virus is stopped by the vaccination. There was never any evidence for this. The available data since summer 2021 have not shown a pattern of the shape that the higher the vaccination rate in a population, the lower the number of infections.[15] The vaccination rate in a population never had any statistically significant meaning for the incidence of infection. There has hence never been any justification for the claim that the unvaccinated are driving the pandemic (e.g. see Kampf 2021). Thus, there was never any scientific basis for discriminating against people based on their vaccination status.

(6) *Overestimation of the scope of modern science*: The relying on poor data, the naïve and uncritical use of models and the crude application of functionalism to human action is accompanied by a breathtaking overestimation of what science can achieve. Modern science has undoubtedly accomplished enormous technological progress. But technological progress cannot be applied to society as "social engineering", with the idea of steering society towards an all-encompassing goal—the "common good"—through state interventions, as if science could bestow on politicians the power to control a virus outbreak.

The mentioned six points show that Ferguson et al.'s work is a textbook piece of political scientism, exhibiting all the shortcomings and mistakes of scientism and its political use that we set out in the two preceding sections. In sum, our assessment is that the corona regime was based on the invention of a postfactual reality. This does not mean that there was not a real danger for some part of the population. But the existence of real danger demands a measured response rather than the spread of fear and

[15] See Subramanian and Kumar (2021), Singanayagam et al. (2022) and Riemersma et al. (2022).

uncontrolled alarmism. Indeed, most of the opponents to the corona regime narrative proposed alternative strategies in consonance with previous data and recommendations. Many of them were already prominent scientists, with proved expertise in epidemiology or public health systems (like, for instance, Martin Kulldorff). What all this suggests is that the "Follow the Science" motto is based on not only a series of methodological and argumentative mistakes but also a postfactual narrative that replaces reality with model simulations for political reasons, to justify massive coercive intervention upon people's lives. It leads to a new, specifically postmodern totalitarianism that, if not stopped in time, will end up in a disaster similar in magnitude to the disasters that the totalitarianisms of the twentieth century caused, as we shall argue in the following.

2.5 The Climate Regime: A Fight Against Windmills

Unfortunately, the damaging relation between science and political power that makes up the corona regime is not an exemption. This regime uses the scientific and political reaction to climate change that has emerged since the 1990s as a template. Many of the elements that enter into the narrative of the corona regime are copied from the climate change narrative: the claim of a scientific consensus, the reliance on flawed or partial models, the crude application of functionalism, the slogan "Follow the Science" to legitimise coercive political measures. There is also here the replacement of empirical research with model simulations that amount to building up a postfactual reality, akin to the six mistakes that we identified in Ferguson et al.'s (2020) paper.

To put it in a nutshell, the climate change narrative runs as follows: humanity faces one of its biggest challenges. The planet is rapidly warming at a rate never seen before. The warming is mostly man-driven, as CO_2 emissions are the principal cause of it. If man-made CO_2 emissions are not rapidly cut off, an apocalyptic future awaits us in a few decades, whose gloomy features can already be seen today. Allegedly, 97% of the climate scientists agree on this (for a critical analysis of how the 97% of consensus is reached, see Tol 2016). The climate change narrative is the justification for many government policies that impose far-reaching coercive measures upon societies controlling what is produced, what is consumed and how it is produced and consumed.

In 2015, in the frame of the Paris Climate Change Conference, the target of keeping global temperatures below 1.5 °C warming was set by political leaders from around the world. The idea was that if the rise of temperatures is not kept below such a threshold, scary climatic consequences will sweep the world more frequently: rising sea levels will swallow entire islands and countries, hurricanes will get more violent and frequent, droughts will be more extensive etc. Climate scientists from universities or governmental agencies mandated by working groups of the United Nations were then asked to resolve what it would take to achieve this agreed target. The reply came in the 2018 special report of the Intergovernmental Panel on Climate Change (IPCC): it will "involve rapid and unprecedented societal transformation" (IPCC 2018, chapter 1.7, p. 77).

However, as with COVID-19, there are actually no reasons for alarmism and panic about climate change that could justify "a rapid and unprecedented societal transformation" imposed through coercion by the state authorities. The presumption of a dire emergency is based on mistakes similar to those of Ferguson et al.'s (2020) work. In the first place, everyone should be suspicious about the claim of a scientific consensus. Modern science was not built upon reaching consensus, but upon a disciplined scepticism and a constant revision of previously held ideas, as we explained in Sect. 2.1.

Claims of a consensus must be taken with care. On what exactly do scientists agree when a consensus is invoked? Climate scientists indeed agree on some specific claims (for instance, that the planet was warming in the last decades), but they disagree on others (the causes of such warming) (see Tol 2016 for further methodological issues). This is particularly important because when consensuses are invoked, these nuances are usually omitted. Furthermore, consensuses cannot serve as *justification* for any scientific claim. As mentioned in Sect. 2.3, eugenics—more precisely, the reasoning leading from the claim of there being people with inferior hereditary traits to their reproduction posing a threat to civilised humanity and from there to the call for political measures—was by and large consensus among biological and social scientists a hundred years ago. However, it would be absurd to consider the eugenic coercive measures to be justified because there was a scientific consensus (or majority agreement) about eugenics at the time. The same goes obviously for any other case of scientism: justification, if there is any, does not derive from consensus or majority view among scientists, neither the justification of scientific theories nor political measures. Any case of scientism is presented to the public as a matter of expert agreement, but it should be obvious that this has no bearing on justification.

Over and above that, the guide to reality is not computer simulations, but observed data. The analysis of the available data does not underpin any alarmism. To mention just two recent examples, in a discussion paper published by the Norwegian central office of statistics, John K. Dagsvik and Sigmund Hov Moen (2023, p. 3) conclude on the basis of the analysis of a large amount of data that “the effect of man-made CO_2 emissions does not appear to be strong enough to cause systematic changes in the temperature fluctuations during the last 200 years” (see also Dagsvik et al. 2020). In a similar vein, Demetris Koutsoyiannis and Christos Vournas (2024, p. 151) state that “the observed increase of the atmospheric CO_2 concentration has not altered, in a discernible manner, the greenhouse effect, which remains dominated by the quantity of water vapour in the atmosphere”.

In the previous section, we pointed out six mistakes on which the scientific justification of the corona regime relies. The same types of mistakes characterise the climate regime: (1) great uncertainty about basic data and climate evolution, (2) underestimation of the limits of modelling, (3) omission of people’s spontaneous adaptation to changes in the climate and the environment, (4) irresponsible and biased assessment of climate change policies and their consequences, (5) reliance upon one particular solution—“vaccinations” against CO_2 emissions—from the outset instead of open-minded research, and (6) a grotesque overestimation of the scope of modern science, as if science were able to steer the evolution of the world’s climate. In the following, our intention only is to substantiate these claims. As with the example of the reaction to the coronavirus waves in the preceding section, we enter into the reaction to climate change only as an illustration of how science exceeds its limits, slipping into scientism and political scientism. This is the target of the book, not COVID-19 or climate change as such.

(1) Climate models and projections are fed with a selection of relevant parameters and initial data for them. However, these models and projections are quite limited due to the enormous complexity of the study object. First, it is well known that many of the global climate changes take time, sometimes millenniums to impose themselves. There were warmer periods that lasted thousands of

years, followed by even longer colder periods (for a reconstruction of quaternary paleoclimate and its climate variations, see Bradley 2015). Why these changes and oscillations occurred and how long it takes to recur depends on many factors, including solar oscillations, volcanic activity and oceanic currents (see Ruddiman 2014; Walzer and Hendel 2023). For instance, it is known that the Atlantic multidecadal oscillations greatly affect the North Atlantic Sea temperatures every 70 to 100 years, while the El Niño-Southern oscillations greatly affect the Pacific Sea temperatures every 4 to 10 years, drastically changing weather patterns worldwide. These factors, and many others, not only occur individually but also overlap.

The evolution of the climate of the Earth is a unique historical process. Since uncountable interactions take place at the same time, it is virtually impossible to get complete knowledge of climate trends by means of computer modelling of the planet's climate via a few parameters and short periods of time. To get a sense of proportion, weather patterns have scales of hundreds of kilometres. However, today's computers cannot even resolve one centimetre if all the potential variables are taken into account. Hence, the selection of the parameters to be taken into account and the fixing of their initial value out of a wide range of possible initial values is inevitable. It is also important to stress that scientists dispose of reliable data about global temperatures only since the 1950s. It is not until 1989 that a global atmospheric record of temperature started to be maintained thanks to a system of satellites, which allows a more fine-grained testing of temperatures' forecasted trends.[16] In any case, in order to assess climate trends that vary in temperatures by about 2–4 °C along several decades, the information available is scarce and imprecise.

(2) There is also a huge underestimation of the limits of modelling in climate science. Once again, computer simulations are taken as representing reality, and their projections as foreseeing the future. However, given the complexity of climate trends and what influences them, it is crucial to drastically simplify the input data and to select just a small number of variables. Climate models necessarily omit many factors to make the models tractable. In particular,

[16] See notably Christy et al. (2007), Maycock et al. (2018), McKitrick and Christy (2020) and Lindzen and Christy (2020).

Richard Lindzen points to relevant facts that actual models are not able to handle well, such as water evaporation, the dynamics of clouds and positive-negative feedbacks. He also stresses that the entire narrative relies on a simple mistake: polar amplification, which neglects that tropic and extra-topic regions exhibit different dynamics (Lindzen 2020a, b). Steven Koonin (2021, chapter 4) offers a clear explanation of the simplifications that climate modelling must introduce.[17]

Once again, there is nothing wrong with omitting information and fixing particular initial values out of a wide range of possible values when modelling. Modern science massively employs abstraction, simplification and idealisation to model and capture phenomena. However, this should be recognised as a limitation when it comes to representing reality and making predictions. It is even more ethically demanding when these projections are used as the basis for policymaking and imposing coercive measures upon people. In any case, it is just delusional to aim to understand and to control the climate of the Earth on the basis of computer models.

(3) Beyond the limits of modelling the climate, most projections of future disasters ignore an essential feature of human behaviour, applying again a crude functionalism to human action. They fail to take into account that people will adapt in ingenious ways to any consequence that changes in the climate could bring about. How people will adapt is unforecastable, but that they spontaneously adapt their behaviour is a fact that has been occurring since human beings appeared on Earth. Thus, even if there is global warming that is heavily influenced by human CO_2 emissions, even if this change causes drastic modifications in the environment, people will adapt. Societies in the past adapted to much colder climates as well as to warmer climates. For instance, at the time of the Roman Empire, Europe experienced considerably warmer temperatures than today. An incipient cold period may have contributed to the fall of the Roman Empire (see Harper 2017).

[17]For open questions, see furthermore, for example, Le Mouël et al. (2019), Connolly et al. (2021) and Richet (2021).

Indeed, there is nothing worrisome about global warming *per se*. CO_2 emissions are not a polluting substance. Burning coal and fossil fuels certainly causes pollution and thereby negative effects on people's health and well-being, but not because doing so emits CO_2. Periods of warming had positive effects on the development of culture and productivity, while colder periods had negative effects. For instance, deaths attributed to cold weather exceed ten times deaths attributed to heat (see Masselot et al. 2023). *There simply is no evidence of a climate change in the present that has essentially only negative effects and that could not be countered by spontaneous and voluntary adaptation, as has always been the case in history.* Today, due to the technological progress in the past two hundred years, we are in a much better position to adapt to climate change than in the past. It is therefore just spreading fear and panic instead of doing science to predict that entire cities will be swallowed by the sea, or that people will die from heat waves without doing anything. A striking proof of adaptation and of how technological and economic progress facilitates adaptation is the fact that the number of deaths due to extreme weather conditions declined by more than 90% over the past century (see Goklany 2011).

(4) As governments were urged to take draconian coercive measures to fight the coronavirus waves, so governments are urged to take drastic coercive measures to reduce CO_2 emissions to fight climate change, too. Of course, if the challenge is such that it can be coped with through spontaneous and voluntary adaptation—as we maintain—all the coercive measures are nothing but pure cost and damage, without any benefit for societies. In any case, the climate regime—that is, comprehensive and coercive regulations by the state authorities to enforce a drastic reduction of CO_2 emissions—has terribly harmful consequences for everyone and in particular for the poorer segments of the world's population (see Shellenberger 2020).

Since the first climate summit in Rio de Janeiro in 1992, governments and environmentalists have exerted pressure to invest in reducing CO_2 emissions by mainly cutting off fossil energies. CO_2 emissions are a by-product of coal and fossil fuels, on which high-developed societies have relied since the early nineteenth century. Exploiting fossil energy is one of the main reasons for the impressive economic growth that was achieved in the last 200 years and from which all segments of the population profited.

Hence, drastically reducing CO_2 emissions through coercive measures amounts to dispensing with the energetic structure that made the world's population much more resilient and much better off.

In the first place, the transition to "green" energy is incredibly costly. After the Paris Agreement in 2015, it was estimated that around 3 trillion US dollars must be invested in "green" energy. Secondly, stringent measures have a huge impact on basic resources on which the worst off mostly depend—food and energy. For instance, the huge growth and support of biofuels at the beginning of the millennium pushed up the overall food prices. It peaked in 2008, throwing millions of people into extreme poverty (e.g. see Nebehay 2008). A new increase in food prices in 2011 had the consequence that more than forty-million persons fell into extreme poverty (see Ivanic et al. 2012). A multi-model assessment carried out by Tomoko Hasegawa et al. (2018) states that stringent climate policies may cause food insecurity worldwide and global hunger.

Climate policies that promote transitioning to "green" energies lead to higher prices because of directing investment to so-called renewable energies, thereby causing a shortfall in the supply of energy. Fossil fuel energy is cheap compared to its alternatives and their level of efficiency and security. Cheap and reliable energy is what has taken people out of poverty. "Green" energy is rather expensive, inefficient and unreliable, frequently requiring fossil-fuel energy as a backup. In our days, "green" energies are able to produce less than 10% of the total energy needed in the United States and around 1% worldwide. In this situation, it is simply a dangerous illusion to imagine a full replacement of fossil fuels by "green" energy.

Energy return on energy input (EROI) is "a means of measuring the quality of various fuels by calculating the ratio between the energy delivered by a particular fuel to society and the energy invested in the capture and delivery of this energy" (Hall et al. 2014, p. 142). Although the calculation can be very complex and society relative, "green" energies generally perform very poorly when compared with fossil energies. Electro-fuels, for instance, have an EROI of around 7, while fossil fuels have one of around 35. One of the consequences of this is that any restriction in the supply of fossil energy and any increase in the demand of "green", expensive energy is poised to push up overall energy prices.

Higher energy prices simply mean that fewer people will be able to afford many essential goods and services. To require poorer countries—and poorer people in general—to substitute directly or indirectly cheap energy with expensive energy is just morally cynical. It is not only a way to hamper their economic growth based on cheap, fossil energy (which took people out of poverty in the United States and Europe in the past); but it is also a way to transfer resources from the poorest to the richest, in the disguise of climate change morality. This transference of resources also occurs diachronically between generations. It is expected that future generations will be richer than present generations. But by drastically reducing CO_2 emissions today, we also reduce economic growth and overall well-being today and thereby cut off the path to future progress. However, pursuing the path of technological progress is what provides societies with the means to adapt to climate change, as mentioned above.

(5) A clear case of counterproductive policies induced by the climate change narrative is Germany. As things stand, nuclear energy is the most efficient and reliable energy source that does not emit CO_2 in its operation. It scores the highest in the energy return on energy input index with an EROI of about 75 (see Weißbach et al. 2013). There is a clear path from wood to coal, from coal to oil and gas and from oil and gas to nuclear energy, in which power density increases (see Smil 2016). This is a path to energy resources that are more efficient, less costly and have less impact on the environment (see Shellenberger 2020). The *Energiewende* that started in 2010 has sent Germany many steps backwards on this path.

To move from nuclear and fossil energies to "renewable" energies not only costs more than 36 billion euros a year but has also terribly harmed the German energy structure. The transition to "green" energies was, at best, faint. Wind and solar power merely represented a meagre 33.3% in 2022 of the German energy mix and only 5% of the baseload,[18] while it only reached 17.2% for the European Union.[19] Fossil energies and notably polluting energies as coal make up for the bulk of the rest, not only without decreasing in the last decades but also increasing since the nuclear plants have been shut down one after another. Coal-fired power plants are put back into operation. The shutting down of nuclear power plants did not speed up the transition to "green" energy, but increased coal mining and thus increased Germany's CO_2 emissions. The *Energiewende* had no effect in reducing Germany's CO_2 emissions in comparison to neighbouring

[18] https://www.statista.com/statistics/736640/energy-mix-germany/

[19] https://ec.europa.eu/eurostat/web/interactive-publications/energy-2023

countries such as France that largely rely on nuclear energy. It has only costs and no benefits. It cuts Germany off from the path to technological progress that will allow to gradually replace coal and fossil fuels with more efficient and less polluting energies without coercive political interventions.

Poor considerations of real costs and energy consumption undermine many of the climate change narrative policies. A clear case are the massive subsidies and coercive regulations that promote electric cars. As mentioned at the end of Sect. 2.3, the European Parliament passed a bill that forbids to manufacture new petrol and diesel cars from 2035 on. According to the climate change narrative, electric cars cause zero CO_2 when driving. Therefore, it seems altruistic and morally desirable to abandon fossil fuelled and polluting cars in favour of cleaner cars (a kind of "luxury belief").[20] However, electric cars are not zero CO_2 emitters all things considered. Their manufacturing largely relies on coal energy, as many cars are manufactured in China, a big coal-burning country. Electric cars' batteries need to be charged frequently, not only adding to the overall demand for energy but also largely relying on fossil energies required to produce the batteries and to charge them. The mining of lithium is not environmentally friendly either. Thus, when all things are considered, electric cars no longer look so clean any more.

When the known harms that such regulations will cause on the overall European economy are added—such as higher energy prices, controls over the labour market, the large subsidies paid with taxpayers' money etc.—regulations are just detrimental. Fixing one strategy of reaction to the challenge of climate change through central state planning in alliance with alleged science undermines the path to technological and economic progress that greatly improved the living conditions in the developed world and that led to a drastic reduction in poverty worldwide over the last decades as well as to energy sources that are more efficient and at the same time less polluting (such as nuclear energy).

(6) Like the corona regime, the climate regime incurs a gross overestimation of what science can achieve, as if politics guided by science could steer the Earth's climate on the basis of computer simulations. The climate regime is another outstanding piece of scientism

[20] https://nypost.com/2019/08/17/luxury-beliefs-are-the-latest-status-symbol-for-rich-americans/

> and its political use. Scientists not only urge populations to go "greener" on the basis of computer-modelled projections but also appallingly censor whoever exercises a dose of scepticism. This does not only mean to replace the exercise of prudence and open-minded enquiry with authority and dogma, but is also a morally reprehensible attitude. In the climate change narrative, scepticism basically means rejection of "The Science", that is, the alleged scientific consensus. But disciplined scepticism has driven modern science from the very beginning. This is another harmful distortion of what science is: the exercise of disciplined scepticism and the promotion of stringent testing of our theories, subjecting them to increasingly rigorous experiments, is rejected in favour of blind, obedient belief. Climate models are limited, and rather poor; the exercise of scepticism, the elaboration of alternative models and the open, free discussion is the only way to improve them.

We thus face, once again, the abandonment of modern science in the very name of science and, consequently, enormous harm to societies. As in the corona regime, which overall resulted only in great damage to people's health instead of protecting them from a threat to their health, the climate regime will only cut us off from the path that hitherto led to a development towards more efficient and less polluting energy sources by cutting us off from the economic means to pursue this development. Germany is about to set the paradigm in that respect: more and real pollution by having to resort to coal as a result of immense investments into unreliable, inefficient and expensive so-called renewable energy resources.

2.6 Intellectual Postmodernism: From the Abandonment of Reason to Wokism

Scientism is an extreme form of scientific realism (see also Peels 2023). Scientific realism can roughly be characterised in terms of the following three tenets: there is a natural world that exists independently of our beliefs about it; our beliefs and scientific theories are true or false depending on the reality of the natural world; science discovers that reality, insofar as we are able to discover it (e.g. see Psillos 1999, introduction). Scientism then takes science to be in the position to answer *all* meaningful questions, insofar as they admit of a rational answer (see above the end of

Sect.2.2). Thus, scientism turns science into the *only* guide to *everything* that is real.

Consequently, it seems that scientism is diametrically opposed to intellectual postmodernism. The former is an extreme form of scientific realism, whereas the latter rejects universal reason and our ability to discover a mind-independent reality through the use of reason. However, as we shall show in this and the next section, the contrary is the case: contemporary scientism meets intellectual postmodernism in replacing reality with its (model) simulation, and contemporary political scientism requires intellectual postmodernism as setting the goals and values (the "common good") towards which social engineering shall be implemented. Hence, again, driving something to its extreme by going from science to political scientism results in its opposite, namely replacing the use of reason with the triumph of unreason.

The term "postmodernism" refers to an overarching cultural movement in art, literature, philosophy and politics that began in the 1960s. Its main characteristic is the rejection of the claim to universal reason that unifies all human beings. More precisely, modernity in general and the age of Enlightenment in particular are characterised by the attempt to use universal reason as a means to limit the exercise of power and authority. As we mentioned before, in science, only evidence and argument count. It is only important *what* someone has to say and completely unimportant *who* says it, that is, what their social status, culture, religion, ethnicity, gender, sexual orientation etc. is. By the same token, the rule of law, implemented in the modern constitutional state, grants the same rights to *every* human being, again independently of their social status, culture, religion, ethnicity, gender, sexual orientation etc.

By contrast, the proponents of postmodernism relativise the use of reason. The use of reason is not universal, but bound to a particular social status, culture, religion, ethnicity, gender, sexual orientation etc. (see Hicks 2004 for a good overview). Characteristics of this intellectual current are, for example, the essay "The Postmodern Condition: A Report on Knowledge" by the French philosopher Jean-François Lyotard (1979) and the book *Philosophy and the Mirror of Nature* by the US philosopher Richard Rorty (1980).

Consequently, the use of reason is reduced to one narrative that competes with many other narratives that are composed by an irreducible plurality of judgements and "phrase regimes" all on equal footing. If modernity aims at unification, postmodernism stresses an irreducible diversity that produces disharmony and confrontation. This abandonment of "big stories"—and notably universal reason—essentially characterises postmodernism. With it, the ideas of progress, objective knowledge, univocity and universality sink down, since they lack any stable foundation. As the postmodernist philosopher Gilles Deleuze paradoxically puts it, the idea is that one "thinks *against* reason" (Deleuze 1962, p. 93, in the English translation 1983). However, the consequence is then that reason can no longer be the means to assess evidence on objective grounds; nor can it be the way to limit power. It can only be the ideology of a specific socio-linguistic community. Reason thus becomes local. Hence, when assessing evidence or seeking the limitation of power, we should ask: *who* is putting forward this reason and *against whom*. This distorts any notion of truth or falsehood, goodness or badness—what is truth or false, good or bad depends on social status, culture, religion, ethnicity, gender, sexual orientation etc.

Besides notorious differences among authors and currents, there are central tenets that characterise this current of postmodernism as opposition against the use of universal reason unifying mankind. In general terms, these tenets point to a form of unrestrained or radical scepticism and to the lack of foundation and dissolution of any form of objective knowledge. To begin, postmodern thinkers tend to endorse some form of anti-realism in the sense that there is no mind-independent reality. Reality is rather fractured, composed of "realities" as social constructs that reflect groups' interests, goals and historical conditions. *Thus, the one reality as the reference and measure of our beliefs and theories is replaced with various simulations*: beliefs and theories do not refer to anything outside of the groups that hold them and hence are not subject to any external standard.

Second, postmodern thinkers promote unrestrained or radical scepticism as a way to undermine any objective knowledge. This leads to particular forms of subjectivism and constructivism, where subjective aspects mediate, or even construct, any form of discourse or intended knowledge. Third, they use genealogy and de-construction as heuristic tools to expose the origin of any knowledge claim as just another narrative or ideology. Knowledge claims are no longer assessed according to standards such as empirical adequacy and logical consistency but are seen as an ideological covering of social groups or historical subjects pursuing their particular interests.

French poststructuralism was probably the main postmodern movement to strongly promote a fragmentation and dissolution of any objective reality in which beliefs and guides to action can be anchored. Each narrative constructs its own reality and canons to judge what is true and what is false. This galloping anti-realism, both in a radicalisation of conventionalism (as in Jacques Derrida and Jean Baudrillard) and the historicisation of knowledge (as in Michel Foucault), led to a form of historical relativism, expressed as a failure in our ability to represent the real and as an irreducible incommensurability among varieties of knowledge.

For the US philosopher Richard Rorty, the heart of postmodernism is the abandonment of any correspondence theory of truth and the acknowledgement that knowledge is essentially fragmented because there is not "One True Account of How Things Really Are" (Rorty 1999, p. 262). To a good extent, the Deleuzian pledge to think against reason can be illustrated through Lyotard's analysis of scientific knowledge. For him, science becomes a form of tribal wisdom with its own myths and legends. This resembles the German nineteenth-century philosopher Friedrich Nietzsche's view of scientific knowledge as chains of useful metaphors, forged into accepted truths. Then, scientific knowledge is nothing but another game, playing with its own rules, object of invention and experimentation for the mere sake of producing new beliefs.

If there is neither a mind-independent reality nor a universal method based on reason to gain scientific knowledge, then modern science loses its rationale. For many postmodern thinkers, the fragmentation of reality and our knowledge of it is not a concern, but a consequence of having abandoned the big narrative of universal reason. Insofar as there is justification at all, it is local and immanent to particular practices and forms of life (see Patton 2001, p. 11875). As regards scientific knowledge, it enjoys a superior status in modernity not because it is capable of formulating theories that are logically coherent and pass the tribunal of experience and reason, but because geographical and contingent details of a historically situated society offer a self-legitimating narrative that puts scientific knowledge as a privileged form of knowledge against others. Ultimately, science is an illusion as others. The results are different forms of constructivism according to which concepts like reality, truth, justice, morality etc. are social, linguistic constructs, driven by diverse groups' interests and aims.

Constructivism and radical scepticism lay out breeding ground for a proliferation of constructivist neo-Marxist theories. Ludwig von Mises (1949, chapter 3) minted the concept "Marxian polylogism" to refer to the Marxist logic of construing any theory (e.g. classical economics) as an ideology, in the sense of being a logical structure that plays the role of justifying interests for the benefit of one particular class. In this sense, there exists a "bourgeois logic" and a "proletariat logic", which differ from one another because they represent different economic class interests. Mises (1949, chapter 3) acutely notes that Marxian polylogism undermines any criteria to judge one logical structure in comparison to other ones: a theory is rejected not because it is false, but because it is the theory that was designed to justify essentially vicious interests, opposed to the genuine, true interests.

Intellectual postmodernism is a radicalisation of Marxian polylogism in that it extends these claims to *any* socio-linguistic group, not only to socio-economic classes. This rationale is assisted with new kinds of analyses of socio-cultural phenomena that seek to uncover the concealed history of concepts, social relationships or theories. The research methods of genealogy and de-constructivism aim to unveil the historical and power-fuelled roots of different logics and rationalities. That is, by explaining how various historical factors converge into a view, postmodern thinkers believe that they explain why the view appears as true, or an institution as justified. Genealogy, for instance, feeds the postmodern idea that objects, concepts and theories were actually socio-historical constructs that express some power imbalance. This leads to consider, for instance, rational argumentation or voluntary market interchanges as the exercise of power and as an intrinsic conflict between parties' interests. What emerges is a perennial history of pervading domination and injustice, where power relations are found everywhere.

However, the genealogical methodology of investigation simply misses the point when it comes to the justification of knowledge claims. Genealogy confuses the context of the discovery of a theory with the context of its justification (see Sect. 2.1 for that distinction). Genealogy suggests that the truth of a theory strongly depends on the details and contingencies that led to the formulation of the theory and its dominance in an epoch or place. Since truth is considered to be historic and contingent, the validity of a theory dissolves itself in local, contingent and arbitrary episodes of particular socio-linguistic communities.

At first sight, it may seem that postmodern intellectuals promote an emancipatory movement across all levels of society. Indeed, many postmodern thinkers believe that postmodernism will liberate societies from the oppression by reason and overarching big narratives. To think against reason is to think against power, and this is to think against an oppressive big narrative. Therefore, postmodernism is often considered to be a way to gain freedom.

However, postmodern radical scepticism towards a mind-independent reality directly undermines objectivity. As we explained in Sect. 2.1, the strive for objectivity in science works as a regulative idea to filter out as much as possible any trace of subjective influence and to open scientific enquiry to everyone regardless of their social status, culture, religion, ethnicity, gender, sexual orientation etc. In this sense, the idea of objectivity means that any claim to knowledge can be put to the scrutiny of anyone with the appropriate skills. One of the most remarkable consequences of objectivity as a regulative idea is that it keeps the exercise of power at bay: it imposes a rational limit to the acceptance of theories or narratives that are guided and justified by subjective traits because they have to face the common tribunal of reason.

Postmodernism views reason and objectivity not as a means to limit the exercise of power and arbitrariness, but as the ideology of powerful groups to impose their values and interests. Objectivity then becomes an illusion, a socio-linguistic mechanism of legitimation, which merely expresses the biographical background and geography of the European modern society. It may well be true that European societies and in particular European elites made reason and objectivity legitimatising mechanisms for the exercise of power; but this, then, is part of the abuse of scientific knowledge, of sliding from science into scientism, by contrast to being a feature of scientific knowledge. Objectivity and reason overall do not as such impose any positive view of reality, but they serve as a *methodology* that sets out the rules to investigate reality: any positive view must face the same tribunal and respect the same rules. But no specific content is prescribed.

Although intellectual postmodernism was originally meant to be an emancipatory and pluralist movement, the radical scepticism and the deconstruction of reality and knowledge deprive individuals of the main means to keep political power and scientism at bay: the strive for objectivity in science and the rule of law as a framework for limiting political power. Unintendedly, intellectual postmodernism opened the way to what it was meant to avoid.

In diluting reality and throwing objectivity away, intellectual postmodernism promotes the view that the representation of reality is not possible. It should be replaced by the idea of *simulation*, that is, a copy or image without reference to the original. For instance, signs (linguistic or otherwise) cannot faithfully represent facts, because there are no facts, at least no facts that are accessible to us through the use of reason. Signs (linguistic or otherwise) are rather images without originals, representations without referents that convey the values of socio-linguistic groups. This idea calls for a radical scepticism with respect to any representation since it is by definition false and suspicious. But this form of radical scepticism and deprivation of any reference to an objective reality is just spinning in the void. In this sense, postmodernism leaves individuals unarmed against power. In the end, power imposes itself. The claim "There is no truth" means in effect "*Mine* is the truth, and *you* shall not question it". By the same token, the statement "You'll own nothing", attributed to the World Economic Forum, means in effect that a certain elite will own everything and its legitimacy shall not be questioned (see Rectenwald 2023, chapter 18).

By way of consequence, paradoxically enough, postmodernism thereby feeds scientism: *contemporary scientism meets intellectual postmodernism in replacing reality with its simulation*, as in the computer model science that is at the basis of the corona and the climate regimes. The corona and the climate change narratives simulate a dire emergency through their (ab)use of computer models. There is no reference to a reality of facts that establish an emergency. Models thus become a science, running unchecked by reality. In this sense, the medical science driving the corona narrative and the physical science driving the climate change narrative are paradigmatic examples of postmodern science—or postmodern distortion of science. *In postmodern science, model simulations usurp the place of reality.* The result then is that the model simulations are instrumentalised as a means to exercise power. This, then, is the contemporary political scientism as exemplified by the corona regime and the climate regime.

Insofar as model simulations are no longer subject to a reality check by exercising methodological scepticism, they are an instance of the general postmodern claim that there is no mind-independent reality that can exert a check on our beliefs, theories and value systems. What is wrong with this is already evident from Mises' (1949, chapter 3) criticism of Marxist polylogism as an antecedent of postmodern thinking. In Marxism, classical economics cannot be either true or false in the usual sense because it is the ideology of the bourgeoisie. It is a theory that is true within the logical structure of the bourgeoisie, but it is false in the logical structure of the proletariat. It is no longer evidence or rational argumentation that proves a theory false, but the logical structure in which it is formulated. According to Marxism, then, the right logical structure is the one that belongs to the historical subject that will realise the socialist society (or, mutatis mutandis, the one that belongs to the right race, the right nation, the right religion, or the right gender etc.). In this sense, there must be theories that are vicious a priori because they are only beneficial to the selfish interest of the class or group that developed them.

The paradox of Marxist polylogism—and of intellectual postmodernism as well—is that there are theories that work in the standard sense of the term (i.e. that are logically coherent and empirically adequate), but that are false because they serve the wrong interests. The paradigmatic Marxist case is classical economics in the eighteenth and nineteenth centuries (e.g. Adam Smith and David Ricardo). According to the Marxist ideology, classical economics provided the bourgeoisie with a rationalisation and moral justification for capitalist exploitation. The defence of free trade by the British economists in the late eighteenth century was, for instance, just the rationalisation of the British capitalists to justify their own greedy motivations and goals. Hence, classical economics can be perfectly true within the British capitalist logic, but it is plainly false in the Marxist analysis: it is the ideology of the class that will be superseded by the proletariat.

By the same token, a physical theory such as Maxwellian electrodynamics could escape, under the Marxist analysis, any relation to truth or falsity. All that can be said about Maxwellian electrodynamics is that it was carried out in the capitalist interest of developing telegraphy. However, this analysis is pointless when it comes to the justification of classical electrodynamics: it may obviously be true that some historical and contingent circumstances led Maxwell and Lorentz to write down the Maxwell-Lorentz equations. But the theory is true or false independently of such historical and contingent circumstances. The justification of a scientific theory must not be confused with the historical conditions that led to its discovery. To mix both and to therefore call them "ideologies" is inappropriate. Mises, for one, points out very clearly the absurd consequences of this view:

> Men use firearms. In order to improve these weapons, they developed the science of ballistics. But, of course, precisely because they were eager to hunt game or to kill one another, a correct ballistics. A merely "ideological" ballistics would not have been of any use. (Mises 1949, p. 77)

For scientific knowledge, the only relevant question is whether a theory can live up to the standards of empirical evidence and rational examination.

Another problem of this methodology, as well as already present in Marxist polylogism, is that it implies an unjustified collectivist analysis. For instance, Marxism argues that classical economists defended free trade because it was beneficial for the British capitalists. However, this is not true as what goes by "British capitalists" involves various individuals, with often divergent interests, that do not respond equally to different incentives and policies. Depending on the conditions, some British capitalists defended free trade policies in some periods, whereas others defended protective tariffs in others. Reality does not simply fit into the analysis because it starts off from the wrong assumptions. Analogously, contemporary emancipatory and post-colonial movements incur the same shortcomings. A genealogical analysis of the historical contingencies for social or scientific theories uncovers an intrinsic struggle where some groups are viewed of as silenced and oppressed. However, these analyses frequently overlook the distinction between the context of discovery and the context of justification as well as commit the fallacy of supposing that there is a unified group that shares common interests and an identity.

If reason is not universal, it can no longer serve as a means to limit the exercise of power. The consequence is that only pure power remains. Facts are no longer facts, but the interpretations of a specific group to gain further power; basic rights are no longer universal, but the ideological framework of a specific group to justify its privileges. Human rights thus are relativised to social status, culture, religion, ethnicity, gender, sexual orientation etc. This means that equal rights no longer apply to all. Certain persons or groups of persons are entitled to more—or fewer—rights than others, depending on their social status, culture, religion, ethnicity, gender, sexual orientation etc.

In this, again, intellectual postmodernism coincides with a form of Marxism that has been widespread since the 1970s, especially in the social sciences, and which makes the following transformation: After the working class, by and large, had not responded to Marxist ideology and blue-collar workers had improved their conditions with the help of voluntary associations such as trade unions as well as cooperatives that provide insurance against ill health, unemployment, old age etc., this new form of Marxism constantly attempts to find other, alleged victim groups who are supposedly oppressed by the modern constitutional state that implements the rule of law. This form of Marxism, also known under the catchword "cultural Marxism", and intellectual postmodernism thus agree in rejecting the way reason is used as a means to limit the exercise of power in science and the rule of law, considering science and the constitutional state as a pure exercise of power by a ruling class—predominantly white, Christian and heterosexual men. "Cultural Marxism" and intellectual postmodernism therefore enter into an alliance that is directed against the claim to universality of the use of reason in science and the rule of law.

In modernity, reason is used to enforce equal rights for all in a legal order and equal validity of evidence and arguments in science, regardless of what the social status, culture, religion, ethnicity, gender, sexual orientation etc. of the persons concerned are. However, if the use of reason is relativised to these factors, then the use of reason is no longer able to limit the exercise of power. *Reason as a means to limit the exercise of power stands and falls with the claim of universality of the use of reason to be the same for all human beings.*

Modernity subordinates power to reason: one must justify every claim to power by giving reasons. The use of reason limits claims to power through the method of scientific research and through the mechanisms of the rule of law. Postmodernism, by contrast, subordinates reason to power: the use of reason is just another, albeit sophisticated, way of exercising power. The consequence of this is that power stands absolute, unsubordinated. Of course, there are emotions that limit the exercise of power independently of reason, but only in the interpersonal sphere and not in the societal or political context. There, emotions can be steered as a means of exercising power.

If the use of reason in the form of the method of putting knowledge claims to scrutiny is no longer respected in science, then science becomes indistinguishable from a state religion imposed on people in order to patronise them: one must "Follow the Science" without being allowed to critically question the knowledge claims and verify them through one's own use of reason. By the same token, if the unconditional respect for the human rights of every person is no longer the basis of political decisions, then the constitutional state slides into an authoritarian state that ends up in totalitarianism: obedience then takes the place of maturity.

If one abandons the basis of modernity of using reason as a means to limit the exercise of power, the result is not that there is no common ground at all. The result is not an anarchy without rules, where everyone lives in their own sphere of facts and with their own moral-normative convictions. There is no such thing where people live together on the same territory. If the basis of modernity, consisting of science and the rule of law, is abandoned, the consequence is not postmodernity in the form of unlimited pluralism, which is often associated with the slogan "Anything goes" of the philosopher Paul Feyerabend (1975). The consequence is the unlimited exercise of power.

This rather long introduction to intellectual postmodernism was important because this intellectual current is at the basis of the emergence of social and cultural movements like wokism. Indeed, intellectual postmodernism is a quite fertile terrain for the development of an ideology that combines many elements that have been popping up in Western society in the last three decades. It views the West as eroded by underlying and out-of-the-view structures that have perpetuated injustices, inequalities and oppression mainly in minority identities. The wokeness regime then promotes the atomisation of society into different groups identified mainly by gender, race, sexual orientation or ethnicity. These groups' interests, the ideology goes, have been historically wronged and oppressed by structures to which we ought to "awake". To uncover and overthrow the oppressive system that preserves various forms of injustices is the main aim of the wokeness movement, paving the way to the cancellation culture, the political correctness, the obedience vows and the moral division between those awake (holding a heightened moral knowledge) and those complicit of the patriarchy, the white race, the unjust capitalism etc.

It is interesting to stress that the roots of the woke ideology came from the social science and humanities departments of first-world universities. Thomas Sowell, probably one of the first to warn us of a growing intellectual movement potentially totalitarian, calls it "the vision of the anointed" (1995). Much water has gone under the bridge since the 1990s, but many elements that Sowell identifies are also part of the wokism movement, mainly in academic environments. For instance, the diagnosis that Western societies are unbearably unjust calls for rectifications that will depend on making people aware of concealed structures and changing their dispositions towards more positive ones. The only way to realise a more equal, inclusive and diverse society is by changing people's minds and attitudes, in the line of the elite of social scientists in wealthy universities. This entails complete disdain to other values (as traditional values) because they are still preyed of oppressive structures and wrong dispositions.

The woke ideology is also mounted on a postfactual reality. It claims that Western societies are infected by systemic racism or patriarchy.

Movements like #BlackLivesMatter and #MeToo have only fed the idea that contemporary Western societies are not much different from Southern United States during Jim Crow Laws or from today's Iran. However, all these claims are factually misguided. Over the past two centuries, Western societies have abolished slavery and segregation based on racial or religious traits and achieved growing levels of tolerance, respect and integration. By failing to acknowledge these achievements, wokism undercuts this path instead of pursuing it. In fact, systematic discrimination in terms of race or gender appears when the woke ideology reigns through its privileging of certain groups and its violation of the principle of equal rights for all. The feminist Pauline Harmange (2020), for instance, goes as far as advocating to hate men.

However, the underlying logic of wokism is invincible: whoever does not see the systematic injustice in our society has not yet "woken up" to the reality of injustice or is a complicit of the injustice. In this way, wokism shields itself from criticism. Woke ideologues may support their claims by evidence. But when evidence is under scrutiny and criticised, the evidence itself is viewed as suspect and ideologically biased because it is inconsistent with the woke ideology. In other words, the ideology is blind to empirical evidence, even though it claims to be based on it.

The most worrisome aspect of wokism is, however, not its distorted view of reality, but that it has descended from the Ivory Tower and entered state and social institutions. Sadly enough, minority identities have been seized by political power to show moral superiority and to run an agenda whose objective is to augment the control of social and even private life without any limits. Legal systems, for instance, are being modified so as to specifically protect minority identities from aggression and alleged discrimination. However noble the intentions may be, the results are clear: unequal treatment before the law, entrenchment of state-granted privileges, comprehensive surveillance of social and even private life and violation of individual rights (freedom of speech, freedom of association, freedom of contracts etc.).

Despite the appearances, wokism is not in conflict with scientism. They rather enter into an alliance in the frame of intellectual postmodernism. On the one hand, scientism, through its abandoning of the standards of science such as, notably, methodological scepticism, lends itself to be used by the wokeness movement as an instrument to debunk the use of universal reason, given that universal reason is abandoned by science itself. On the other hand, wokism delivers the norms in the guise of the goals and values—the "common good"—towards which science shall steer society. Thus, *contemporary political scientism as exemplified by the corona and the climate regimes is the outcome of the alliance between scientism and intellectual postmodernism*, as we shall explain in the next section.

2.7 Political Scientism Today: From Intellectual to Actually Existing Postmodernism

With the corona regime, the climate regime and the wokeness regime, we undergo the transition from intellectual to actually existing postmodernism, ushering in a new, specifically postmodern totalitarianism. This transition parallels the transition from intellectual socialism—that is, Marxism—to actually existing socialism, as implemented, notably, in the Soviet empire and in mainland China. Intellectual socialism envisages the liberation from coercion in the guise of the communist society in which the state will have disappeared. The idea is that once all sorts of oppression are removed, everybody contributes according to their abilities, and, *as a consequence*, the needs of everybody will be satisfied—as in Marx's slogan "From each according to his abilities, to each according to his needs!".[21] By the same token, as explained in the previous section, intellectual postmodernism envisages a liberation from the power that certain groups exercise notably under the pretext of reason, science and the constitutional state. The idea is that once this exercise of power is removed, the result will be something analogous to the communist society in that there are no longer any social inequalities.

In both cases, the transition to the ideal society requires the creation of a new human being. That new human being is, however, not only utopian, but its very conception is contradictory. As regards communism, the problem is that the capacities of humans are limited, whereas their needs are unlimited. As regards postmodernism, the problem is that any voluntary interactions among humans will result in inequalities, precisely because the capacities of human beings are different and the circumstances in which social interactions take place are various and contingent and beyond our control. Nonetheless, voluntary interactions, as they occur in free-market societies, are such that all participants profit from them: each of the

[21] "Jeder nach seinen Fähigkeiten, jedem nach seinen Bedürfnissen!" in the German original (*Kritik des Gothaer Programms* 1875, quoted from Marx and Engels 1962, p. 21). English translation quoted from Marx and Engels (1966, p. 19).

participants is better off as a result of voluntary interactions—otherwise, they would in general not agree to the interaction and it would hence not be voluntary.

Both intellectual socialism and intellectual postmodernism advocate employing coercion in order to realise the ideal society—that is, coercion without any principled limits, since what is at stake is the highest good, namely the realisation of the ideal society. Since the very conception of that society is contradictory, the result is only coercion and thus the opposite of the envisaged end. Thereby, we get to the characteristic features of actually existing socialism that are reproduced in actually existing postmodernism. Both implement a regime of encompassing regulation of social and even private life that leads only to destruction. In the latter case, it is the destruction of health in the corona regime, the destruction of the natural environment in the climate regime through cutting off the path to technological progress and the destruction of social relations in the wokeness regime. Indeed, actually existing postmodernism is the old wine of actually existing socialism in new bottles. Socialism did not disappear from the political scene with the demise of the Soviet empire (see Niemietz 2019 on the multiple transformations of socialism). Actually existing postmodernism is by far the most influential and dangerous resurrection of actually existing socialism since the fall of the Soviet empire.

Michael Rectenwald (2021) hits the nail on the head in attributing the corona politics and its drive through alleged science to the mode of operation of postmodernism. As he puts it, "without objective criteria, there is no court of appeal other than power" (Rectenwald 2021). Rectenwald (2018) diagnoses the transition to "practical postmodernism" (pp. xiii, 114–117) in terms of the development of wokeness and cancel culture in US universities, which amounts to pure tyranny. "Practical postmodernism" expresses the same idea that we name "actually existing postmodernism" in reference to what used to be known as actually existing socialism in contrast to the intellectual doctrine of Marxism.

Indeed, what has happened since spring 2020 has much deeper roots than merely being an exaggerated reaction to a virus wave that served some special interests of scientists, the pharmaceutical industry, the media and politicians to the detriment of the vast majority of the people. The corona regime constitutes a complete break with the fundamentals of modernity through which Western societies obtained enormous gains in the protection of rights and improvements in the overall standard of living for all segments of the population. Since 2020, the breaking of this trend has not only drastically restricted human rights and largely paralysed economic and social life but also resulted in parts of science abandoning the path of open-ended, truth-oriented research in the service of political activism. The control mechanisms of the rule of law to limit the power of governments—especially the courts up to the constitutional courts—failed. *If a virus outbreak on the scale of the coronavirus waves can have such consequences, then any event can lead to abandoning science and the rule of law.* That is the danger that is inherent in what has happened since spring 2020.

In particular, the central legal principle of the presumption of innocence was put upside down, thus reversing the burden of proof. The mechanism that led to this development is the following. Coronavirus waves are real and put some people at risk, including causing deaths. But their danger is grossly exaggerated through irresponsible scientific modelling—the invention of a postfactual reality—so that they appear as a life-threatening danger for the general population. This not only provokes mass hysteria but also makes it possible to portray everyday activities—like going to work, meeting friends or even venturing outdoors for a solitary walk—as endangering the well-being of others. Hence, instead of citing concrete and verifiable evidence that someone endangers others by pursuing a particular activity, everyone is put under general suspicion, having to prove in advance that one is innocent and does not risk endangering others. Since, however, that risk never is completely absent, regulation and surveillance can grow without limit and penetrate far into private life.

Yet, this is not an isolated episode. It is staged against a pre-existing background. Technological progress and an overall improvement of health conditions in Western societies led to an increased sensibility to our well-being as well as to the natural environment. There is also more sensibility to injustices, arbitrary inequalities, cruelty and punishment. In consequence, societies have developed different mechanisms to produce consumable goods in an environmentally friendlier manner, to enforce cooperative and peaceful social relationships, to change daily practices that improved hygiene and thereby health etc. Against this background, which is just one of the consequences of better living conditions, it is, however, possible to develop various mechanisms of social control that take advantage of this sensitivity. The risks of being infected by a virus, or of being harmed by climate change, are then grossly exaggerated to control people's lives.

By reversing the burden of proof, the intellectuals and scientists that promote the corona, the climate and the wokeness regime put unbearable responsibilities on individuals. These responsibilities go far beyond what individuals can directly or significantly influence within their respective scope of action. After all, any free action, or the omission of an action, can be construed as somehow contributing to the spread of a virus, to climate change or to perpetuating historical injustices simply by belonging to a certain ethnic or other social group.

As mentioned in the previous section, Marxism sought to mobilise the blue-collar workers, contending that they were the victims of the capitalist reason that led to the modern constitutional state. When this attempt failed, the current known as "cultural Marxism" sought to find all sorts of alleged victim groups of capitalism and the modern constitutional state. With the corona regime, the climate regime and wokism, this mechanism is pushed to its extremes: we are all social victims at least of the spread of a virus or of climate change. That is to say: we are all perpetrators. That is the reason why viral waves, as the coronavirus waves, as well as climate change, are selected among the various challenges that we face: these challenges make it possible to fabricate a postfactual reality in which we are all perpetrators. Any of our free actions can be portrayed as contributing to the spread of a virus, or as contributing to climate change. As regards wokism, it comes in as making it possible to stage-specific groups as perpetrators, in particular, white, Christian and heterosexual men. Due to contingent historical circumstances, these groups were the main pillars of science and the rule of law, paving the path to scientific, technological and economic progress from which all segments of the population profited. Now, wokism picks out any of their actions or views as perpetuating a historical injustice towards some other minority groups.

This is how intellectual postmodernism ushers in the transition from scientism to political scientism: scientism as such is purely technocratic. It does not provide any norm or aim that gives the necessary guide to technocratic reasoning. Going from scientism to political scientism, it is not clear what to "Follow the Science" shall imply and with respect to which goal "social engineering" shall be done. Intellectual postmodernism fills this lacuna, prescribing what "Follow the Science" and "social engineering" shall aim at by fabricating narratives that turn us all into perpetrators at least with respect to spreading viruses or contributing to climate change. A regime of comprehensive social control can subsequently be installed in these respects that then is legitimised by alleged science.

However, what takes in effect place is an illustration of what we described in the preceding section in terms of the abandonment of reason by intellectual postmodernism ending in the unlimited exercise of power: an elite imposes their goals and values on society as a whole without giving normative reasons for these goals and values, but by presenting them as alleged scientific facts. Political scientism then means that the social engineers claim for themselves a freedom to set goals and values that they deny to the people they subject to social engineering. It is appalling to see how some scientists estimate that their scientific qualification gives them the moral authority to steer other people's ways of life. It is just plain arrogance on the part of some scientists to think that people who do not have the same mathematical-technical intelligence as they cannot act in a responsible way and therefore have to submit to scientists dictating them their lifestyle. In the same way as the religious fanatic only recognises the elect as fully fledged human beings, so the scientist who promotes political scientism only sees him- or herself and his or her own kind as fully-fledged human beings. From this perspective, other people are mere objects whose lives the scientist engaged in political scientism takes him- or herself to be entitled to steer in virtue of his or her scientific competence.

Scientists engaged in scientism and its political application, therefore, destroy science and its position in society in three ways: (i) Firstly, they destroy the objectivity of science: the corresponding knowledge claims no longer refer to facts; as norms, goals or values are presupposed, they no longer abstract from the subjective perspective. (ii) Secondly, they destroy the scientific method: one can no longer apply the method of disciplined scepticism to a knowledge claim that is passed off as an advice for coercive political actions. (iii) Finally, they destroy the social acceptance of science: those people who have other goals and values than the ones supposedly prescribed by science will be portrayed as unscientific—and are in consequence likely to turn against that sort of science, which they are nevertheless supposed to fund through their taxpayers' money. The recognition that science enjoys in society across all different value systems depends on its objectivity. If science turns into a moral guideline, we fall back into the pre-scientific age.

This is indeed what happens: science takes the place of a religion employed by political power to install a regime of comprehensive social control. People have to cleanse themselves from allegedly endangering others through their everyday activities by submitting to political authorities and their experts. This amounts to comprehensive central planning of the lives of individuals and their social communities such as families by central state organs or supra-national bodies in conjunction with massive interventions upon economical activities. The submission to this arbitrary power generates high imbalances in societies as authorities determine which individuals are to be favoured in contracts simply because they belong to certain groups (ethnicity, gender etc.) in order to compensate for historical injustices, how one may move around, what one may eat and how one may heat one's home in order to stop climate change and in what form one may have social contacts in order to prevent the spread of viruses.

The general result then is the following. Individuals can only free themselves from the general suspicion of endangering others through their everyday activities by acquiring a state-issued certificate or state-enforced social pass by which they clear themselves. In the corona regime, this takes the form of a negative test, a vaccination certificate or generally a social passport. In the climate regime, this may, for instance, take the form of tracking one's carbon footprint, reducing air travel or acquiring an electric car. In the wokeness regime, this may, for instance, take the form of "diversity and inclusion vows" in university job applications. Generally speaking, in this way, *the licensed person takes the place of the responsible citizen.*

These regimes destroy the basis of liberal democracies. First of all, human rights become privileges granted by a central authority. Free individuals are thus replaced with persons who acquire rights from political authorities by acquiescing in the conditions that these authorities set for the exercise of these rights. This mirrors the operation modus of totalitarian regimes, in which freedom is gained by submitting to the regime's orders. Getting vaccinated to cleanse oneself of the suspicion of spreading the coronavirus, going "green", becoming awake from concealed networks of pervading injustices are the way to gain freedom rights as a privilege granted by the political authorities and to avoid ostracism and cancellation. This is obviously not about freedom or rights but about privileges as a reward for conformity to the orders of a regime that is totalitarian in that it claims the permission to regulate any aspect of social and even private life.

Furthermore, the social and political mechanisms to keep the exercise of state power checked are thus undermined. One of the aims of liberal democracies is to avoid arbitrariness by the state's institutions. However, the conditions under which a state authority grants social passes or credit are arbitrary. There never is zero risk in social interactions. Risks are inherent to life; it is not possible for a government to regulate risks in a way that is binding on all without sliding into arbitrariness. Risk assessments are subjective, depending on individuals' preferences and contexts. Hence, such regulations cannot avoid favouring the preferences and risk assessments of certain people at the expense of the preferences and risk assessments of others. Their negative effects and potential non-compliance then trigger further regulations and controls. But this only reinforces the effect that people are disadvantaged by growing regulations. Therefore, all these governmental policies ultimately have the opposite effect of what they claim to aim at. The so-called coronavirus protection measures provide ample illustrative material for this, as explained in Sect. 2.4, as does, for instance, the German *Energiewende* (see the end of Sect. 2.5).

To disguise the arbitrariness of these regulations, they are then embedded in an Orwellian reinterpretation of language, which also resembles totalitarian practices. In the corona regime, the meaning of central epidemiological terms such as "pandemics" and "infection" is changed (see Sect. 2.4). More general terms also suffer such a semantic transformation. "Solidarity", for instance, no longer means working to ensure that others can lead a self-determined life as best as they can under the conditions that are specific to them. "Solidarity" now means imposing a certain way of life on everyone, regardless of whether there is any demonstrable benefit. Thus, one exhibits "solidarity" with one's elderly relatives by cutting them off from all direct social contact and even letting them die in isolation—a cruelty that echoes the crimes committed by past totalitarian regimes in the name of science and morality. Freedom as self-determination is branded as egoism and the aforementioned reversal of the relationship between human rights and the state is presented as responsible freedom: the licensed human being who submits to the collective guided by a cast of science-priests and politicians is portrayed as a free human being—for example, in the German book with the language-twisting title "libertarian authoritarianism" by Carolin Amlinger and Oliver Nachtwey (2022).

The reversal of the presumption of innocence, the transformation of rights into privileges, the vision of the state as the grantor of rights (instead of their protector) and the explicit arbitrariness are elements that indeed lead to a new form of totalitarianism. To illustrate this transition, let us briefly consider two examples. After depicting dire climatic cataclysms, the 2019 report by the American Association for the Advancement of Science, "How We Respond", reads:

> While these problems pose numerous risks to society and the planet, undoubtedly the biggest risk would be to do nothing. Science tells us that the sooner we respond to climate change, the lower the risk and the costs will be in the future. (American Association for the Advancement of Science 2019, p. 2)

In this document, all the fearmongering and alarmism as well as the morally reprehensible call for coercive political actions are exemplified. While the benefits (if any) of these actions are uncertain, it is from the outset certain that they will cause much harm (as pointed out in Sect. 2.5). This document is one illustration, among many others, of the political scientism that drives the climate regime.

As an even more striking example of political scientism and its moral reprehensibility, consider the "7th Ad Hoc Statement on the Coronavirus Pandemic" of the German National Academy of Sciences Leopoldina issued on 8 December 2020. It reads:

> Despite the prospect of an early start of the vaccination campaign, it is *absolutely necessary from a scientific point of view to* quickly and drastically reduce the still clearly too high number of new infections through a hard lockdown. (Leopoldina 2020a, our translation and emphasis)

At least the following six points are noteworthy about this document:

(1) It claims to represent *the* scientific point of view without discussion and without a plurality of voices presented with reason within science.
(2) It ignores the state of research on lockdowns. By the end of 2020, numerous studies were already available citing evidence that lockdowns cause at least more harm than good, if they have any beneficiary effects at all (e.g. see Sect. 2.4 and the references in footnotes 13 and 14 of this chapter).
(3) It suggests that the spread of the coronavirus can be controlled by a hard lockdown of a few weeks, namely during the Christmas holidays of 2020/2021, although it has been clear since the study by Ferguson et al. from March 2020 that lockdowns are a matter of several months.
(4) The evidence for the effectiveness of a hard lockdown is based on a model simulation that relies on a single example, namely a lockdown

in Ireland. However, as is well known, individual examples can be found for any claim—certainly, there are also individual examples of men biting dogs.

(5) Coercive political measures that involve massive violations of human rights, as well as causing great harm including deaths, are presented as "absolutely necessary from a scientific point of view", as if there were no dividing line between science and normative proclamations.

(6) Given that by December 2020, there was a controversy about such coercive political measures in the general public and in the political parties, the National Academy interprets its mission statement that it "champions the freedom and appreciation of science", "promotes a scientifically enlightened society and the responsible application of scientific insight for the benefit of humankind and the natural world" and is "an advocate of human rights" (Leopoldina 2020b) as implying that it is the Academy's task to help out the government: science becomes the *ultima ratio* to justify central planning of people's lives—including their social contacts and family life—together with the suspension of fundamental rights guaranteed by the constitution.

In short, if one looks for a paradigm example of political scientism, one can hardly find a better document than this statement by the German National Academy of Sciences.

With this document in her hands, chancellor Angela Merkel justified a hard lockdown in the German parliament just one day later, on 9 December 2020, by invoking laws of nature such as the law of gravity and the speed of light (Merkel 2020). This is the most momentous political consequence of this perversion of science: no one is responsible for laws of nature such as the law of gravity. We discover laws of nature. We have no influence on them. If coercive political measures that are contrary to human rights are entailed by natural laws, then no one is responsible for these measures and their consequences—neither the scientists who recommend these measures (they only discover facts that follow from natural laws) nor the politicians who implement them (they only follow "The Science"). There is no choice, at least no rational choice between different courses of action.

As we pointed out before, turning science into a political programme to control people's lives means going down a road that ends up in totalitarianism, namely in an unrestricted regulation of people's lives justified by alleged scientific findings. These alleged scientific findings, given the apparent urgency of the situation, must be immediately and directly translated into political action, irrespective of human rights. *Political scientism uses science as a weapon against human rights and against the control mechanisms of the constitutional state.* The aforementioned statement by the German National Academy thus provides ample illustrative material for how charlatanism can be passed off as science. Much worse: If one seeks to understand how science can become the stooge of a regime taking on totalitarian traits, one does not have to revisit history. One only needs to consider statements issued by erstwhile venerable scientific institutions such as the quoted one and their political use.[22]

Totalitarianism is not necessarily a regime that asserts itself through the use of physical violence. Physical violence usually comes into play only when the population no longer believes in the narrative on which the regime in question is based. Totalitarianism is characterised by unlimited social control and unlimited regulation of people's lives by a political authority. German Chancellor Olaf Scholz's statement that he does not recognise any red lines in the fight against the coronavirus waves through coercive political measures[23] is an expression of such a totalitarian attitude. In a democratic constitutional state, such a statement, if not corrected immediately, should lead directly to impeachment.

There are numerous features that the new postmodern regimes have in common with previous totalitarian regimes. First, there is a group of experts who present themselves in the media as the mouthpiece of science and who claim to have the knowledge of the common good. In other words, these experts claim not only specialised knowledge of matters of fact but also moral knowledge. These experts' vision usually comes with a crude functionalist and technocratic view of human beings, considering them as mere objects that can be manipulated and steered towards a scientifically devised common good. People are diminished, lacking the knowledge and the moral qualifications to achieve the common good.

Totalitarian regimes exhibited similar practices in the past. As mentioned in Sect. 2.3, communism relied on alleged scientific laws of social

[22] Personal note by Michael Esfeld: Being myself a fellow of the Academy, I have first asked the President to revoke this statement in an internal letter (without receiving an answer) before publicly denouncing it.

[23] https://www.tagesschau.de/inland/scholz-spiegel-corona-101.html, 12 December 2021.

development to tailor society and to move towards the deterministic aim of a classless society. Massive violations of human rights were not only justified by relying on these laws but also presented as inevitable. This took the form of, for instance, direct action against social groups and persons branded as class enemies, up to and including their outright murder. Since political authorities were merely instruments of scientific laws of social development (analogous to laws of nature), they remained unaccountable for their action, bearing no personal responsibility for the harm caused. National socialism, by the same token, invoked biological laws about races to steer society towards a pure-blooded society. Once again, the path to this ultimate goal included the physical extermination of allegedly inferior races. Again, the national socialist leaders in the German Third Reich were convinced that they only carry out what was predetermined by the laws of nature.

An important difference between the past totalitarian regimes and the new postmodern ones is that, today, the common good towards which society is steered is no longer an absolute good presented as the ultimate goal of history. This would be part of a big narrative that is absent in intellectual postmodernism. If anything is absolute in today's totalitarianism, it is an absolute evil, namely the extinction of humanity by a killer virus, by man-made climate change or whatever other catastrophe scenarios one may come up with and build into supposedly scientific models. The view of the common good enters then as the means to avoid an apocalypse. But this is not an absolute good as the risks are various and changing: health risks, climate risks, emotional risks etc.

Underlying the narrative of apocalypses and omnipresent risks is an attitude formulated by the philosopher Hans Jonas in his book *The Imperative of Responsibility* (1979, English translation 1984). Jonas' imperative of responsibility is that scientists consciously and deliberately develop models that highlight apocalyptic scenarios, and that society and politics direct their actions to prevent these scenarios from occurring. This imperative has shaped an entire generation of scientists and social leaders. This is, however, the opposite of an imperative of responsibility: it is irresponsible to set apocalyptic scenarios absolute, as if they were scientific knowledge about the future, and to direct all action towards avoiding their occurrence. It is clear from the outset that doing so causes in any case great harm and violates people's basic rights, whereas the benefits, if any, are unclear. The corona and the climate regime with the harm that they cause for public health and the natural environment provide ample evidence of this. Such an imperative of responsibility thus turns out to be yet another instance of the reinterpretation of concepts mentioned above: a call to morally reprehensible action under the pretext of moral concepts with positive connotations. In setting possible apocalyptic scenarios absolute, the issue is not protection from real risks, but staging risks as a pretext to build up a comprehensive system of social control.

At this point, we come full circle to the mechanism described above: everyday actions are placed under general suspicion of harming others. The most radical way then to avoid risks and harming is by the creation of new human beings, as attempted by the totalitarian regimes of the twentieth century. Those who ideologically advance these regimes aim to create a new human being who is purified from the spread of harmful viruses, from climate-damaging behaviour through energy consumption, from past injustices contributing to diversity, equality and inclusion. But this evidently collides with freedom as self-determination: the new human being is, in turn, a completely designed human being. This entails the destruction of not only independent thought and action but also creativity and innovation.

Along with the blueprints of a new human being comes the undermining of social units that organise social and economic life independently from political authorities and self-proclaimed experts. Charity organisations, mutual aid communities, religious communities and small- and medium-sized businesses are overtaken by central planning. For instance, big companies can be centrally controlled and bribed through state benefits. A clear case is the benefits granted to "climate-friendly" big companies to obtain contracts, investments or loans, even if they are not economically competitive. This business model was taken to the extreme with the coronavirus vaccines: exemption of liability for damages, so that profits can flow without economic risk; in addition, artificially created demand through guaranteed purchase of the products through government vaccination programmes, without any proof of their effectiveness and safety according to scientific standards. Universities are forced to comply with regulations that promote woke values, diminishing a more comprehensive representation of viewpoints at classes, or self-governance by university bodies. Above all, these totalitarian practices destroy families. This is not a surprise since families are the first obstacle to schemes for central control of social processes. The reason is that families are *par excellence* autonomous decision-making units, which are incompatible with third-party decision-making by central political authorities or scientific "experts".

After the devastating consequences of the religious wars, the hallmark of modernity is the absence of a collectivist narrative that holds society together. The absence of such a narrative provokes counter currents to the overriding trends such as communism and national socialism. These postulate a final goal of history, namely the classless or the pure-blooded society. These big collectivist narratives differ from the pre-modern state religion in that the final state lies in this world and is to be realised here and now through coercive political measures, which include the extermination of individuals and entire groups who are seen as standing in the way of the realisation of the final state.

Postmodernism replaces these big narratives with many small, interchangeable narratives that take the place of the rule of law as the bond that holds society together. Thus, certain facts—such as the coronavirus waves, climate change, historical injustices—are selected. Each of them is knitted into a narrative that presents these facts as an all-encompassing reality that has a moral-normative status. Although this basis is factual, one thereby constructs a *postfactual* social reality: there are coronavirus waves, but there is no new virus that is dangerous to the general population. There is climate change; man-made CO_2 emissions arguably are one factor among several that influence climate change; but there is no evidence of a man-made climate change that threatens the survival of humanity and that therefore cannot be dealt with through adaptation and technological progress, as has always been the case. There have historically been pervasive injustices and discrimination against parts of the population on the basis of ethnicity, religion, gender, sexual orientation etc. But today, there is no such systematic discrimination in Western societies—except for the discrimination that the postmodern regimes themselves introduce by excluding certain groups as impure, like the non-vaccinated in the corona regime.

These narratives have the purpose of leading us from an open to a closed society: the basis of society is no longer a legal order that safeguards human rights and within which different forms of life can flourish. This order is replaced with a collectivist narrative that distributes privileges and that orients society towards a common goal with all economic, social and ultimately individual life being directed to that goal. What is specifically postmodern about this is that there is not one narrative that postulates an absolute common good, but a plurality of small narratives each of which postulates a partial common good and constructs a postfactual reality that replaces facts with simulations in that respect.

Ayn Rand is credited with the statement that one can ignore reality but cannot ignore the consequences of ignoring reality.[24] The postmodern regimes have understood the lesson of this quote: the regimes of communism and national socialism collapsed when their narratives could no longer ignore the consequences of ignoring reality. The postfactual construction of the corona regime also ignored reality, in which the obsessive focus on the coronavirus waves put aside other health risks and the consequences of the coercive measures. But the postmodern regime does not end there. It can quickly set up another, equally postfactual narrative that ignores reality. The new regime can in turn be allowed to run until the consequences of having ignored reality can no longer be ignored—and then move on to the next narrative. This is what makes the postmodern regime so dangerous: it is by no means finished by debunking the narrative that has just prevailed (like the corona narrative). It is then simply followed by the next narrative. That is why it is so important to reveal the common basis of all these narratives.

[24] The evidence is the essay "The Objectivist Ethics", chapter 1 in Rand (1964). However, the statement is not a literal quotation from this essay.

CHAPTER 3

How to Restore Science

3.1 How to Overcome Scientism and Its Political Use

In the previous chapter, we have shown that modernity faced the branching of two paths: one leading to modern science and the other one to scientism. We have also explored their foundations and worked out how scientism and actually existing postmodernism have taken over the political and cultural stage nowadays. This take-over is wrapped in small narratives such as the corona, the climate change and the woke narratives. However, it would be a mistake to assume that to overcome actually existing postmodernism, it is sufficient to leave each of its regimes and narratives behind. The corona regime has by now largely faded away. But its underlying scheme persists with the climate and the wokeness regimes. Even if these regimes were removed, we would have to expect that actually existing postmodernism will fabricate yet other narratives to continue totalitarian social control. It is therefore indispensable to dismantle scientism and its political use at its roots. We show how to do so in this chapter. The first two sections lay out practical steps that everybody can implement by employing their capacities of judgement. The last two sections provide the intellectual foundation to remove scientism by rehearsing the principles and the rationale of modern science as set out by Descartes. Kant and subsequently the linguistic and social-normative turn

M. Esfeld, C. Lopez, *Restoring Science and the Rule of Law*, Palgrave Studies in Austrian Economics,
https://doi.org/10.1007/978-3-031-71186-2_3

in 20th philosophy as well as Austrian economics from Hayek and Mises to Rothbard and Hoppe show us how to develop a science of human freedom that complements natural science and that allows it to prosper by keeping it within its limits. Restoring science, then, requires both reviving natural science as set out by Descartes and accomplishing a human and social science based on human freedom.

To debunk scientism and its political use, one only has to bear in mind the fact that humans continuously adapt to new challenges and that they do so in numerous and unpredictable ways. History shows that the knowledge acquired through spontaneous adaptation always beats central planning. Since humans first set foot on Earth, they have faced innumerable lethal threats. Extreme poverty, massive famines, deadly diseases, hostile natural elements were common currency in past societies. Nonetheless, our contemporary societies have been incredibly successful in dealing with all these threats. It was rapidly understood that technological innovation and specialised knowledge (based on reason and evidence) were the spearhead for making us better off. But it is not until the eighteenth century that the quality of life started to skyrocket almost everywhere on Earth: high technological innovation, scientific knowledge, high specialisation in production and an economy based on a strong division of labour and free trade have made us much better off in more or less all aspects of life. The human ability to cooperate, to work peacefully together in an economy of division of labour and free, voluntary interchange not only creates immense wealth but also allows us to overcome threats more easily.

In parallel with this history of astonishing success, there is the history of doomsday prophecies. But this is rather a history of systematic failure. Although life conditions have become much better, they may become worse in the future if a particular trend continues. On this basis, prophets of doom paint various scenarios in which we are always on the brink of jumping off the cliff. The basic problem of most of the gloomy predictions is that they simply assume that a particular trend will go on as it is, without people adapting and being innovative in various and unpredictable ways.

The following example can be instructive. During the nineteenth century, mobility in big cities such as London and New York greatly depended on horses. The transportation of goods and people was possible thanks to horse-powered buses and carts. It is estimated that by the late nineteenth century, London's streets were populated by more than 50,000 horses. Horse-powered transportation was one of the causes of population and city size growth, along with an increase of overall quality of life. People could move faster from one place to another, goods were delivered more rapidly, facilitating and increasing production. The more horses there are, the better this is for the economy of a city. Nonetheless, urban planners predicted an around-the-corner emergency and a sudden reversal of the city's path to more prosperity. In 1894, the *Times* published an article in which it was foreseen that in 50 years, London was going to be buried in horse manure, if the horse-powered transportation trend went on. The present may be bright, but the future is poised to be buried under more than three meters of horse manure on the streets. That was *the great horse manure crisis* of 1894 (see Burrows and Wallace 1998).

The problem was deemed intractable by urban planners of the day. Around 50,000 horses produced 570,000 kilograms of horse manure to be wiped daily. If the trend went on, human capacity to regularly clean the streets would promptly going to be insufficient. Therefore, an urban catastrophe was looming. It was just a matter of following the trend. And yet, cities were not buried in manure and the problem solved itself (see Davies 2004). The commercial introduction of motor vehicles in the early twentieth century rapidly replaced horses. The less horses there were, the less manure they produced. The crisis thus vanished. Two things are noteworthy. First, the dire predictions were off target as they did not take into account people's response to incentives, people's adaptation and imagination. Second, no urban planner, no government intervention, no expert was part of the solution. Technological innovation, inventors and entrepreneurs—such as Karl Benz, Gottlieb Daimler and Henry Ford—within a largely free, open society as cities in the United States and Europe were in the late nineteenth century made the problem fade away.

As another example, let us come back to eugenics. As mentioned in Chap. 2, Sect. 2.3, in the 1920s, "The Science" was presented to the public as the doomsday prediction of the demise of civilised humanity as a result of the trend that people with inferior hereditary traits tend to have more offspring than people with superior hereditary traits. The past may have seen the rise of civilised humanity, but it is poised to vanish due to this trend. This prediction was then employed to legitimise forced sterilisations of people with allegedly inferior hereditary traits. After the national socialist regime had driven eugenic politics to the extreme, the eugenics movement faded away also in Western democracies in the period after the Second World War. If there ever had been a real problem underlying the apocalyptic prediction of the demise of civilised humanity as a result of the reproduction of people with inferior hereditary traits, that problem simply vanished, because that trend did not continue. Instead, pursuing the path of scientific, technological and economic progress led to improving the living conditions of all segments of the population, including in particular the worst off. As a result, there was no longer a trend of significantly different reproduction rates that could be attributed to different segments of the population. The interventions by scientists and state officials, overriding the basic rights of their victims, contributed nothing to this. They were simply morally reprehensible, if not outright criminal.

The same applies to the doomsday predictions about the health effects of the coronavirus waves. To postulate a continuing trend of exponentially increasing infections is not science, but shows a lack of imagination, if not outright stupidity. Basing coercive political measures that override fundamental human rights on the postulate of such a trend is, again, morally reprehensible, if not outright criminal. What was effective in dealing with the coronavirus waves was the spontaneous adaptation of people's behaviour and natural immunity. The state interventions in the name of "The Science" caused only great harm to public health and far beyond, as documented in Chap. 2, Sect. 2.4. The same applies to climate change: insofar as there is a problem, we have today much better means to adapt to changes in the Earth's climate than previous generations had. Simply pursuing the path of scientific, technological and economic progress leads in itself to the development of ever more efficacious technologies that have less impact on the environment, as is illustrated by the path from burning wood to burning coal, oil, and gas to using nuclear power as energy source.

Thus, when scientists make such doomsday predictions and call for state interventions in the name of "The Science", the appropriate reaction is to simply laugh at them. One should remember the success of spontaneous adaption to new threats in the past and how this spontaneous adaption regularly trumped central intervention. Scientists that predict an upcoming catastrophe do not only ignore history but also treat their fellow citizens with disdain in considering them to be unable to adapt intelligently to a new challenge. Societies have faced innumerable problems along their history, and they will face innumerable problems in the future. The good and old news is that people adapt and invent new strategies to overcome issues. And the good and not that old news is that open societies are in the best position to deal with problems and crises: a system of free exchange under the division of labour and free interchange of ideas, goods and services, within a general legal framework that protects basic individual rights, has the best chances to not only overcome crises but also still push societies' wealth forward. Our overall message thus is simple: the best thing that we can do is to pursue the path of scientific, technological and economic progress, because this path puts us in the best position to cope with the challenges that the prophets of doom seek to exploit to hold power over us.

3.2 How to Escape the Dominion by "Experts" and Strengthen One's Judgement

Political scientism promotes "experts" as an elite that guides people's lives. In order to resist and become immune to the sirens of political scientism, one has to strengthen one's capacity of judgement. Indeed, a failure of judgement has guided the political responses to the coronavirus waves in particular. The faculty of judgement is the ability to assess a situation appropriately in its context and to respond to it proportionately. In Kantian terms, judgement brings intuition—the perception of a given situation—together with concepts in the sense of the classification of that situation. The assessment that the coronavirus waves pose an extraordinary danger to the general population and that the only possible response consists in coercive political measures are an unprecedented failure of judgement among large sections of the elite of scientists, politicians and media communicators, and subsequently also among vast parts of the population that trust this elite.

But what would the application of judgement look like? One should first ask oneself the following question: *Why should it be that there is an identifiable group of experts—medical professionals in this case—who have a privileged knowledge of how to deal with a new situation that is not available to the general public?* Of course, psychologically it is tempting to believe that there are such experts. For their part, it is tempting for these experts to claim such knowledge for themselves. Instead of succumbing to this temptation, however, one should ask oneself whether there are good reasons for this assumption.

We know of health hazards whose effects do not become apparent immediately, but only years or decades later. Radioactivity is an example. However, the following can also be argued in this case: one feels the effects of a significant dose of radioactivity immediately. Therefore, caution is advisable even with small doses of radioactivity. Be that as it may, the crucial point in our context is that a virus outbreak does not fall into this category. In a virus outbreak, everyone sees the consequences immediately and adjusts their behaviour accordingly. If the virus is dangerous, then one sees people going about their normal lives getting infected, becoming seriously ill and possibly even dying. If one does not see such consequences in one's environment, then there is no dangerous virus circulating there. But, more crucially, only individuals know what the right response to a virus outbreak in different circumstances is. This is so not only because risk assessment is highly subjective but also because the socio-economic circumstances greatly vary. Finally, if it is a novel virus, there is no privileged expert knowledge that lay people cannot verify by observing their surroundings.

Nevertheless, there is something very useful that experts like scientists, decision-makers like politicians and opinion-makers like journalists can do in such a situation: they can work to avoid panic reactions and maintain the conditions for a sober evaluation about what is really happening. However, many scientists, politicians and media professionals did exactly the opposite: they deliberately spread panic, among other things with arbitrary model calculations and computer simulations that painted disaster scenarios on the wall without being supported by real-world evidence.

Let us consider the question of privileged expert knowledge. We can subdivide this question in the following way:

i. *Existence* of a strategy that achieves the common good: *Why should there be exactly one strategy to respond to a new situation that is appropriate for all individuals, societies and cultural and economic contexts?*
ii. *Knowledge* of such a strategy: Even if such a super-strategy for the common good were to exist, this question arises: *Why should politicians and their advisors be trusted to have privileged knowledge of the best strategy that is superior to the knowledge of individuals, their choices and their specific circumstances?* Coercion in the form of overriding human rights presupposes that whoever is exerting coercion has valid reasons for believing that people would not arrive at the behaviour that coercion is intended to achieve through the use of their own reason. Thus, in other words, why is it assumed that a privileged group has knowledge that enables them to centrally select a strategy for all from above, rather than learning from below, through trial and error, which is the best strategy for dealing with a new situation?

iii. *Ability and moral qualification* to implement such a strategy: Even assuming that such knowledge exists in identifiable individuals—politicians, scientists—the following question arises: *Why should politicians be trusted to implement a best strategy for the common good, rather than succumbing to multiple special interests of particular groups* (all of which, of course, are touted as a common good)?

When a viral outbreak strikes, one spontaneously adapts one's behaviour to the given situation without needing a state authority. State intervention could be needed at most to facilitate people of certain professions to voluntarily go about their work. The German lawyer Carlos A. Gebauer (2021) sums up this fact succinctly:

> The difference between an actually proven pandemic and one that is only discursively assumed becomes apparent in the state's legislative response. In a merely assumed pandemic, citizens are forbidden to leave their homes. In an actual pandemic, by contrast, they would have to be ordered to go to work depending on their profession. (our translation)

In other words, if the state locks people up in their homes and forbids them to pursue their social contacts, then one can immediately infer that the claim of an extraordinary, dangerous situation that requires rapid state action overriding basic rights is unfounded. If such behaviour were really appropriate to the situation, people would develop it on their own.

Just as in the case of coercive political measures, so also in the case of the vaccination campaign the core of the issue is obvious to everyone without expert knowledge: if vaccination is available, then individuals afraid to be infected will voluntarily be vaccinated. Then, they will be protected from infection and from passing the virus on. The non-vaccinated decide according to their risk assessment to do without the protection of vaccination and bear the risk of infection with negative consequences for their health themselves. They cannot put others at risk against their will, if everyone has been offered the possibility of protection through vaccination. If this is not so, and it is claimed that everyone, regardless of their will, must be vaccinated because protection is not achieved otherwise, it simply follows that the medical treatment advertised as a vaccination does not meet the standards of a vaccination.

The same assessment that Carlos A. Gebauer formulates with regard to lockdowns therefore applies to the vaccination campaign: if vaccination against a viral outbreak is available, then government coercion is at most necessary to make the vaccine available first to those most at risk and to prevent everyone else from being vaccinated. This was indeed the case in spring 2021 when the aim was to ensure that those at risk were given the opportunity to be vaccinated. But when the state authorities resort to coercion to enforce a certain vaccination rate on the entire population—again, overriding basic human rights—one knows that the narrative of a universally dangerous virus and a vaccination preventing its spread is a fake.

Hence, to put it in a nutshell, we do not need a vaccination campaign against the coronavirus. We need a vaccination, so to speak, against letting us deprive of our faculty of judgement. It is advisable to listen to the information that experts give, but not to blindly follow instructions from them without checking what they recommend against our own perception of the situation and our own preferences, values and risk assessments. Moreover, we should never hesitate to apply the—scientific—method of scepticism: always ask critical questions. Nobody is an expert because they are presented as an expert in the media or have government approved diploma. As always, the proof of the pudding is in the eating. Someone is an expert—or a medical doctor, a scientist, a teacher, a preceptor, a lawyer or a pastor for that matter—to the extent that they are recognised as such by the people to whom they offer their services on the basis of the people's own judgements.

3.3 Restoring the Fundament: Human Freedom as the Presupposition of Science

There are thus two steps that everybody can take in order to overcome the regimes of political scientism. The one step is to insist that the path of scientific, technological and economic progress is the way to cope with new challenges, while political interventionism guided by alleged science has always only resulted in great harm. The other step is to sharpen one's judgement to escape from the dominion by "experts". However, this is not sufficient to restore science. Restoring science, in the modern sense, is the only way to completely escape from scientism and its political use in the postmodern narratives. Doing so requires putting forward a positive and constructive alternative to the way that leads from science to scientism. To achieve this, we have to go back to Descartes.

As explained in Chap. 2, Sect. 2.1, Descartes conceives modern natural science with the aim of making us masters and owners of nature ("nous rendre comme maîtres et possesseurs de la nature", Descartes 1902, p. 62). For science to be able to be employed in improving our living conditions through technological progress, it has to be objective, abstracting from our subjective perspective on the world. Everything else then follows from objectivity being the hallmark of modern science. In particular, it follows that the method of science has to be the one of disciplined scepticism, submitting all claims to objective knowledge to a rigorous scrutiny. Moreover, *it follows from science being objective that it meets a principled limit in everything subjective, that is, in particular, human thought and action.*

If the scientific standpoint consists in stepping back from all subjective elements, then the consequence is that one cannot grasp the subjective features within this standpoint. Hence, modern natural science cannot, as a matter of principle, grasp its own presuppositions in human thought and action. However, if modern natural science cannot grasp human thought and action, what can do so? And how can one justify normative statements if science, in the sense of modern natural science, cannot do so? Providing a positive and constructive answer to these questions is crucial in order to cut off the fallacious way from science to scientism and from there to political scientism. Unless there is such a firmly established answer, scientism will step in, suggesting that the success story of modern science can seamlessly be continued by taking it to capture human thought and action as well.

In other words, there is nothing wrong with modern science as such. Restoring science does not require touching the fundamentals that Descartes has laid out. All that we need to restore science and to continue its success story instead of pursuing its self-destruction through scientism and its political use is a clear and firm answer to the two mentioned questions.

The main step to answer these questions is to move on from René Descartes to Immanuel Kant. Kant elaborates on how to grasp human freedom and thereby provides the alternative to scientism with the framework of modern philosophy set by Descartes. Kant writes in the *Prolegomena to every future metaphysics* (1783):

> If an appearance is given to us, we are still completely free as to how we want to judge things from it. (*Prolegomena* § 13, note III; quoted from Kant 2002, p. 85)

Indeed, freedom consists in the first place in the formation of beliefs, thoughts or judgements. If, for example, one has the sense impression of a stick that *appears as* broken on the surface of the water, one may be led to judge that the stick *is* broken. But the sense impression does not impose this judgement. The person can take a position with respect to the sense impression and form a judgement. Judgements contain concepts; but concepts do not follow from sense impressions. We are free to form concepts based on sense impressions and whatever else may be given to the mind. In this vein, the philosopher Wilfrid Sellars (1956) forcefully denounces what he calls "the myth of the given": nothing that is given to the mind can as such impose a thought on a person or justify the thought.

The same applies to actions. There is no automatism from sense impressions and desires to actions. In an extension of Kant's quote, one can say: if a desire is given to us, we are still completely free as to how we want to act. The person can position herself with respect to her desires and form an intention to act. Intentions presuppose judgements and thus concepts. They are not a mere reaction to desires. Intentions are a judgement about how the world *should be*, combined with an attempt to bring about the corresponding change in the world through a physical movement.

Freedom in forming judgements about the world (freedom of thought) and freedom in forming intentions to act (freedom of the will) are one and the same freedom: if a person has the freedom to form judgements about the world, then she also has the freedom to position herself with respect to her own stance in the world; this is the freedom to form goals and intentions to act. When biologists, neuroscientists or psychologists uncover factors that influence our actions, it does not follow that we do not have free will. The moment we recognise these factors, we can position ourselves in relation to them and change our actions (see Kotchoubey 2012; Mele 2014; Esfeld 2020, chapter 3). Therefore, cognition liberates. It frees us from factors that have previously influenced our thoughts and actions without us being aware of them. Once we have unravelled them, however, they enter the deliberation about our thoughts and actions: we can position ourselves in relation to everything we are aware of and are therefore free to do what we want with it.

The philosopher John McDowell, being part of what is known as the Pittsburgh school in philosophy going back to Wilfrid Sellars, illustrates this point by asking what it would take for a wolf to think and act:

> A rational wolf would be able to let his mind roam over possibilities of behaviour other than what comes naturally to wolves. … [This] reflects a deep connection between reason and freedom: we cannot make sense of a creature's acquiring reason unless it has genuinely alternative possibilities of action, over which its thought can play. … A possessor of *logos* cannot be just a knower, but must be an agent too; and we cannot make sense of *logos* as manifesting itself in agency without seeing it as selecting between options, rather than simply going along with what is going to happen anyway. This is to represent freedom of action as inextricably connected with a freedom that is essential to conceptual thought. (McDowell 1995, § 3)

This entails two things. First, descriptions about wolves as a species (a scientific description, for instance) cannot *justify* anything about what rational individuals do. Second, such descriptions enter as information to be considered rationally, that is, from a critical stance. If rational beings could not step back and ask for the epistemic credentials of, for instance, scientific descriptions, then they would not be rational beings anymore. McDowell's ingenious comparison sharply draws the line between—scientific—descriptions of facts and normative attitudes. If we are genuinely rational, we can also ask for the epistemic credentials of any description of facts that is given to us.

Indeed, *freedom, reason and normativity form an inseparable trio*. For thoughts and actions—and only for these—one can demand and cite reasons. This is what the ancient Greek expression *logon didonai*, giving reason, means. Events in nature happen according to certain regularities (laws of nature). One cannot demand reasons for them. If a storm knocks down a tree and someone is harmed as a result, one cannot demand reasons and thus no accountability for this event. The same applies to the behaviour of animals: if a cat plays with a mouse before killing and eating it, one cannot demand reasons and thus accountability for the cat's behaviour. By contrast, humans can position themselves with respect to their sense impressions, desires, needs, instincts etc.; they are therefore free in their thoughts and actions and hence accessible to reasons and thus accountability. Normativity comes with this: if a person positions herself with respect to her sense impressions, desires, needs, instincts etc., then the question is what the person *should* think and do, that is, what is right in thinking and acting and what is not.

The justification of a thought or an action can therefore not consist in the reference to individual sense impressions or desires. The sense impression of the stick being broken on the surface of the water can incite the thought that the stick is broken; but it does not justify this thought. The justification consists in placing the sense impression in a context with other sense impressions and forming thoughts on this basis. It then turns out that the sense impression of the stick being broken on the surface of the water does not fit in with other sense impressions, such as touching the stick. Thus, a sense impression, for example, is classified as reliable because it is supported by other sense impressions. But the classification as reliable is a normative, evaluative assessment. In this way, the person builds up a reasoning context. However, this does not mean that justification is explicit all the time when one thinks and acts. The claim is only that it makes sense to ask for justifications when thinking and acting (see Brandom 1994 for a detailed account in the vein of the Pittsburgh school).

By the "we" in the quote above, Kant means persons as transcendental subjects—that is, as beings whose behaviour does not simply occur according to natural laws. After the linguistic turn in philosophy in the twentieth century, we understand the transcendental subject today, with good reason, as a person who uses language. The crucial point is that we can form judgements only together, only in social interaction, by correcting one another mutually and thereby giving precise meaning to the terms we use, as Ludwig Wittgenstein in particular points out in his *Philosophical Investigations* (1953, §§ 138–242). Thinking and acting is bound to language, and language is a social affair. Thinking and acting thus take place in a social context.

It is not the case that individuals each form thoughts and intentions for action on their own and then enter into social relations with other individuals depending on whether and to what extent they find such interactions expedient. Social relations are primordial and not derived from isolated individuals. However, social relations do not determine a person's thoughts and actions. Rather, it is by forming language that we collectively free ourselves from biological constraints, thereby achieving individual freedom. Through using language and forming concepts, it is possible for each person to position herself with respect to her sense impressions, needs and desires and thereby build thoughts and intentions for action. In short, the social nature of human beings enables individual freedom as self-determination of one's own thoughts and actions (see Esfeld 2020, chapter 3, for a detailed account).

In this way, social relations, including language use, are based on normative attitudes. Each person adopts the normative attitude in relation to other persons to judge their utterances and actions as correct or incorrect and is aware that the other persons adopt the same normative attitude to her. Again, this happens implicitly rather than explicitly. However, a situation can arise at any time in which the normative attitude becomes explicit in the form of the question of justification. This is why freedom and normativity are the two sides of the same coin: because we adopt normative attitudes, because it makes sense to ask for justifications in thoughts and actions, we are free. And because we are free, because we can position ourselves with respect to our sense impressions, needs and desires, we ask ourselves what we should think and do and adopt normative attitudes towards each other in our social interactions.

This is why considering a being as a person is not an empirical description of its behaviour, but the expression of a normative attitude towards it. This attitude can be successful or unsuccessful, depending on whether the being towards which one adopts a normative attitude reciprocates this attitude or not. When a normative attitude is not reciprocated, there is no inter-personal relationship. For instance, a robot is merely a machine unable to normatively reciprocate our attitude towards it; artificial intelligence may exist, but machines do not think and act (see Landgrebe and Smith 2022). When it comes to living beings of the human species, by contrast, this attitude is reciprocated. Every human being, including babies and small children, has the potential to reciprocate this attitude. Wilfrid Sellars aptly writes:

> To think of a featherless biped as a person is to think of it as a being with which one is bound up in a network of rights and duties. From this point of view, the irreducibility of the personal is the irreducibility of the 'ought' to the 'is'. ... To think thoughts of this kind is not to *classify* or *explain*, but to *rehearse an intention*. (Sellars 1962, section VII)

From this fundamental normativity derives a moral realm of rights and obligations. In adopting a normative attitude towards a being, one recognises this being as enjoying the freedom to think and act. To put it differently, to grant the right to exercise this freedom in the form of own formation of thoughts and actions *is* to recognise someone as a person, normatively. Moral obligations and rights (as freedom rights) are normative conclusions drawn from a normative premise. To deny people these rights is to refuse to regard them as persons, not to adopt a normative attitude towards them.

Classifying a being as a person hence is not a scientific description of the being in question. Granting a being the status of a person is a normative expression of the recognition of its freedom. The recognition of a being as a person therefore implies, in Kant's terms, the moral obligation to always treat this being as an end in itself and never merely as a means to an end (see, for instance, *Groundwork of the metaphysics of morals* 1785, section II, in Kant 1911, p. 429/English translation Kant 1996, p. 80). The justification of this obligation depends only on what all human beings have in common as characterised by the inseparable trio of freedom, reason and normativity. Since what they have in common is something normative, there is no naturalistic fallacy involved here, that is, no derivation of an *ought* from an *is*.

Consequently, when recognising the distinction between facts and norms, we do not only recognise the "off-limit" signs to science and insulate a normative realm but also stress where and how science oversteps itself. We emphasised repeatedly that the pillar of modern science is the strive for objectivity, for looking at the world from a point of view from nowhere. But this already supposes the distinction between facts and norms; it supposes that it is possible to look at the world objectively, abstracting from all subjective and normative traces. The rise of modern science is a crucial episode in our history because the world could be looked at as deprived of meaning and subjectivity. McDowell (1994, lecture IV) calls modern science the "disenchanted conception of nature", that is, a realm of laws of motion in contrast with the realm of reasons. Hence, the distinction between facts and norms does not only limit modern science's scope but also makes its rise possible in the history of Western thought.

Events in nature happen according to certain regularities (laws of nature). To capture them in an objective theory of the world, they must be left disenchanted, that is, deprived of meaning, subjectivity and norms. It is then pointless to demand reasons from nature or factual descriptions: a world deprived of reasons and normative features can in principle not justify reasons or normative discourses. To do otherwise is just to incur in the well-known naturalistic fallacy. Any fact-based description of the world in scientific terms is objective, that is, non-evaluative. It merely says what *is* the case. From non-evaluative, scientific premises, we can only move to non-evaluative conclusions, that is, to other fact-based descriptions. But, to use fact-based descriptions of the world to justify normative attitudes is to move from non-evaluative premises to an *evaluative* conclusion. This is the is-ought gap, which lays at the core of the naturalistic fallacy. Of course, hidden evaluative premises can always be brought up, but it must be clear that the evaluative premises, not the non-evaluative ones, are doing the justificatory work.

The recognition of the is-ought gap is the reason why irrational animals are not accountable for their behaviour (as when a cat kills a mouse): they just do what comes naturally to them. They cannot give reasons for their behaviour because they do not have the options to let their minds roam over alternatives—they do what animals instinctively do. But persons (as McDowell's rational wolf) can step back, reflect and take a critical stance to what comes naturally to them, to what is given to their mind. They are therefore free in their thoughts and actions and thus accessible to reasons and thus accountable for their thoughts and actions.

Against this background, it is evident why the "Follow the Science" slogan is so deeply wrong. Modern science is about facts, about objective, disenchanted descriptions of the world. Modern science gives us information about the world and thereby enhances our possibilities for actions. Thanks to scientific knowledge, we can "roam over" more options. But it is up to us which judgements we form on this basis, which possibilities we embrace. Scientific knowledge can enhance and enrich us, but how it does so depends on our aims and values, on which scientific knowledge has nothing to say. Hence, "The Science" cannot impose any action upon us.

One can also make use of one's freedom in the full positive sense, with its connection to rationality and good reasons, by organising one's life in a self-determined way in renouncing on the use of scientific knowledge. This is what religious communities such as the Amish do when they refuse to use modern technology. Their members lead a self-determined life precisely because, in their view, such a life requires them to reject the conveniences of modern technology and rely solely on the knowledge and abilities of common sense. In this sense, consequently, the connection between science and freedom as self-determination is *contingent*.

However, the more successful science is in opening up new options for fathoming a self-determined life, the stronger the temptation becomes to present this connection as *necessary* and thereby to slide from science into scientism. Science then appears as a condition for developing freedom as self-determination. Then one has to "Follow the Science" in order to lead a self-determined life. This, then, makes us dependent on "the experts" who have the relevant knowledge—or at least can credibly pretend to have it. It therefore amounts to putting science above freedom as self-determination and ends up in the self-destruction of both science and freedom. Science cannot itself justify the normative "Follow the Science" slogan. To put it differently, we cannot follow the science to justify "Follow the Science": the justification of this slogan requires placing it in a normative network that lies beyond the scope of modern science. However, doing so then makes evident why this slogan is nonsensical.

The same point can be made more generally. We have said that there are clear off-limits signs to modern science. A normative realm in which freedom unfolds is off its limits. But it is important to realise that freedom in forming judgements is also a *prerequisite* for modern science. That is why scientism goes astray: to formulate and accept scientism as an attitude is a judgement that persons can form and for which one can demand reasons. The *object* of scientism—matter in motion, including the matter of our bodies and brains—could not justify the *theory* that everything that exists is matter in motion and can accordingly be grasped by the method of the natural sciences. Only a person can justify a theory in the exercise of her freedom of judgement. Every theory is formulated, accepted and justified in a normative framework by claiming the freedom to form a judgement on the basis of what is given (sense impressions, data, observations etc.). In other words, justification is only possible by claiming the freedom to be able to position oneself in relation to everything that is given and to form a judgement by oneself. Thus, scientism claims freedom for the formulation and justification of this stance while at the same time negating this freedom. *Consequently, freedom in thought and action neither require a scientific justification, nor could natural science provide such a justification. Quite to the contrary, the very possibility of doing science presupposes freedom in thought and action.*

3.4 Kant, Mises, Rothbard: Two Types of Science and the Transcendental Argument

In light of the last section, we can say that there are actually two types of science: on the one hand, there is science in the sense of *modern natural science,* which is widely regarded today as the sole science. This is the science built upon the strive for objectivity (freed from subjective influences). It is largely empirical and a posteriori: observation and experiments are the standards against which scientific claims are assessed. Its methodology is disciplined scepticism, consisting in the strict scrutiny of factual statements by trying to find evidence that would prove the corresponding assertion to be false.

On the other hand, there is another type of science that refers to the *preconditions* for discovering facts in the world by addressing what empirical science refrains from, namely our subjectivity. This science works with what is known as a *transcendental argument* in philosophy. Unlike arguments that refer to transcendent objects that are supposed to exist beyond space and time, a transcendental argument refers back to the subject itself, namely to the conditions of the possibility for the person to recognise objects in the world and to change the world through actions. These are conditions whose negation would consist in a performative contradiction: the performance—the assertion—of the negation of these conditions would itself be an act that makes use of them.

Descartes formulates a transcendental argument in the *Meditations on First Philosophy* (1641, second meditation): it is impossible to deny that I think. For if I deny that I think, then the performance (act) of denying is itself an instance of thinking. It follows that I exist as long as I think. This is a transcendental argument that establishes the existence of something—one's thinking—without depending on observations and without being able to be confirmed or refuted by observations. A science that works with transcendental arguments is therefore not empirical, but an a priori science. Its propositions cannot be falsified by empirical methods. However, logical errors—errors of reasoning—can occur in the formulation of a transcendental argument. Hence, methodological scepticism is also appropriate towards transcendental arguments: they can be falsified by exposing errors of reasoning.

Descartes does not explicitly have freedom and normativity in mind in his transcendental argument about thought. Therefore, a substantive theory of thought and action cannot be built upon the Cartesian argument. But Kant does so. Freedom and normativity in thinking cannot be denied without committing a performative contradiction. The act of contesting would itself be an act that claims freedom in forming a judgement (instead of being, for instance, mere noisy sounds) and that makes a claim to validity. The same applies to actions: one cannot deny freedom in forming intentions to act without the act of denial itself being an exercise of that freedom. This act is an action—at least a speech action—for which it makes sense to demand reasons. Consequently, it is not mere behaviour, but the exercise of the person's freedom to position herself in relation to sense impressions, needs, desires etc.

This is another way to see positively the limits of natural science. There is another science built upon transcendental arguments that can grasp thoughts and intentions to act. In order to recognise that a being has sense experience and to gain access to the qualitative content of its experience, one must take one's own sensory experience as a starting point and attribute sensory qualities to other beings in analogy to one's own experience. That is to say: one must not abstract from one's own subjective perspective. Nevertheless, the statements about the sense experience of other beings can then be objective in the sense that they are true or false depending on what sense impressions these beings actually have—or had, in the case of historical research; but the formulation of such statements precisely requires not to abstract from the first-person perspective.

Thus, the central feature of the humanities and social sciences is not the "disenchantment" of its subject matter, but the contrary: it needs to assume that the regularities and objects it deals with are subjective at its roots. Hayek illustrates this necessary presupposition by the problem that an archaeologist faces when trying to determine whether what looks like a rock is an artefact made by humans or a natural product. For archaeology to be a social science aiming to describe and discover past ways of life, one has to understand "the working of the mind of prehistoric man" (Hayek 1952, p. 27); and for this, the archaeologist has to assume some knowledge of the workings of the human mind from inside. In other words, it would be impossible to formulate a social science "from the point of view of nowhere" since the scientist would then be incapable of distinguishing a mere rock from a stone-made artefact. The same goes for other social sciences and humanities as a whole, in which the interpretation of regularities depends on assuming that purposeful human action creates such regularities, and not nature. This interpretation would be impossible without assuming that they were produced by individuals with a mind like ours.

This does not mean that the hypotheses and theories of the humanities and social sciences, the *humanities* as distinct from *science*, are not objective. But it must be clear that their objectivity is different from that of the natural sciences: it is not achieved by abstracting from the subjective perspective. Quite to the contrary, it is obtained by starting from the subjective perspective, namely by attributing experiences, thoughts and intentions to other persons in analogy to one's own experiences, thoughts and intentions. Thus, one relates the object of research to one's own experiences, thoughts and intentions. Natural science and humanities differ from each other in that the objectivity of the latter is objectivity on a normative foundation, namely by adopting a normative attitude towards the object of study. The archaeologist who studies a stone-made artefact establishes a *normative* attitude to the object, in which interpretation of meaning, decisions, purposeful actions and aims intervene. Even if one then acknowledges that the other people and cultures studied have different normative attitudes than one's own, one's interaction with other people and cultures is still normative. The crucial point to understand the difference between natural sciences and the humanities and social sciences is then to recognise that the latter are mainly about normative attitudes rather than mere behaviour. This is only possible from a perspective that itself includes a normative attitude.

This idea is also at the roots of the development of modern economics, mainly in the so-called Austrian School of Economics. For instance, classical economists such as David Ricardo, Thomas Malthus and Adam Smith held the so-called labour theory of value according to which the value of a good was determined by the labour time employed to produce it (or by the quality of land). This delivered, among other things, the idea that there exist "natural prices" towards which market prices ought to tend, which provided a strong connection with the sort of objectivity of the natural sciences. Also, they believed that the whole economic system ultimately relies on Divine plans that guarantee the stability of societal regularities.

However, economics towards the late nineteenth century turned to a more subjective approach, the *marginal revolution* by William S. Jevons, Carl Menger and Leon Walras, in which individuals' assessments and their interactions became central. Marginalism starts from the premise that there is a distinction between total utility and marginal utility. Individuals, then, rank successive units of goods (and services). This ranking is necessarily subjective since each individual has different preferences. This subjectivism then implies that the value of various goods and services is strictly dependent on individuals' preferences. Then, the opportunity cost will vary with decision-makers' ranking and evaluation of alternatives. This explains the illusion of neo-classical economists when relying on concepts such that of "equilibrium"—equilibrium can only be achieved if all individuals share the same knowledge, rank their preferences equally and no disturbing factors intervene.

The idea of a normative-subjective foundation of economics has led to understanding economics as a science in which what human beings believe about what things are and what goals and values they set for their actions is the study subject. That's why Hayek maintains that

> it is probably no exaggeration to say that every important advance in economic theory during the last hundred years was a further step in the consistent application of subjectivism. That the objects of economic activity cannot be defined in objective terms but only with reference to a human purpose goes without saying. (Hayek 1952, p. 31)

This assessment is at the basis of Hayek's theory of spontaneous and unintended consequences, which are the object of the social sciences: they explain the emergence of this unintended order from human actions and beliefs. It is not the case that the social order is spontaneously generated by natural, or irrational forces that are aliens to us; quite the contrary, the social order is spontaneously generated by human action, by the purposeful actions of human beings as us. But to understand this, it is necessary to understand the normative-subjective foundation of this knowledge and how it differs from the natural sciences.

The most radical and clearest formulation of this insight is praxeology, Ludwig von Mises' (1949) theory of human action. Mises formulates a logic of action that is a priori. He explains the contrast to the natural sciences, but also to classical economics, in these terms:

> External objects are as such only phenomena of the physical universe and the subject matter of the natural sciences. It is human meaning and action which transform them into means. Praxeology does not deal with the external world, but with man's conduct with regard to it. Praxeological reality is not the physical universe, but man's conscious reaction to the given state of this universe. Economics is not about things and tangible material objects; it is about men, their meanings and actions. Goods, commodities, and wealth and all the other notions of conduct are not elements of nature; they are elements of human meaning and conduct (Mises 1949, p. 92)

Since the foundation of economics is human action, all that is needed for it is knowledge of the essence of human action, which is knowledge that we all possess because we are acting human beings. Once again, praxeology is in turn based on the subjective-normative foundation of a particular kind of knowledge, as we developed it in the previous section. Thus, Mises starts out from the statement that all human beings act in a purposefully way. This statement cannot be denied without committing a performative contradiction. Mises then attempts to deduce the theorems of economic science from this axiom, since economic relations are just various modes of human action.

Mises' praxeology is an outstanding example of how one can build a social science on an a priori rather than on an empirical basis. His work receives little attention in economics because academic economists are predominantly oriented towards adopting the method of the natural sciences and thus towards adopting assumptions that bring economics closer to the natural sciences. This can be seen, for instance, in the criticisms that the Austrian School raises against neo-classical economics. There are many methodological and technical differences between the two. What is crucial in our context is that the Austrian School recognises the distinction between different kinds of sciences and how important it is to respect the subjective-normative foundations of knowledge to understand human behaviour. While neo-classical economists are prone to assume constant rationality, objective scale of preferences, homogeneous tendency to maximisation and cost-benefit analysis as well as complete, objective and constant information by all economical agents, Austrian economists emphasise the fragmentation of knowledge, the subjective nature of preferences and value, and divergent reactions that individuals have in the face of unexpected circumstances.

Despite the popularity of neo-classical economics in mainstream academic circles, Austrian economics has been integrated to different degrees in many areas of analysis and policy making, such as subjective elements in the price system. Moreover, praxeology is currently further developed. To mention a few examples, Jesús Huerta Soto (2008) argues against neo-classical economics and the idea of a price equilibrium. Walter Block (1986) thoroughly analyses many societal and cultural issues from a praxeological approach. In Germany, Andreas Tiedtke (2021) recently published a detailed account of praxeology as a guide to action based on the distinction between non-aggressive and aggressive behaviour and Thorsten Polleit (2022b) elaborates on this approach in terms of a priori, transcendental arguments applied to economics.

However, Mises' praxeology is a descriptive enterprise. It does not take into account the point driven home by Kant (transcendental subject) and

Sellars (social-normative, linguistic turn) according to which being a person is not a matter of fact—be it a matter of fact seized by an empirical, *a posteriori* science such as modern natural science or a matter of fact seized by an a priori science such as praxeology—but a normative status such that the recognition of a being as a person consists in adopting a normative attitude to this being. In short, an a priori science of thought and action is not a descriptive science, but a science that presupposes a normative attitude to its subject matter.

That is why the work of Murray Rothbard is of crucial importance for the further philosophical development of the Austrian school. *Rothbard implements a normative turn in Austrian thought*: he places Mises' praxeology within the framework of a normative theory of freedom. Rothbard expresses this normative turn in the preface to *The Ethics of Liberty* (1982):

> I at no time believed that value-free analysis or economics or utilitarianism (the standard social philosophy of economics) can ever suffice to establish the case for liberty. Economics can help supply much of the *data* for a libertarian position, but it cannot establish that political philosophy itself. Political judgments are necessarily value judgments, political philosophy is therefore necessarily *ethical*, and hence a positive ethical system must be set forth to establish the case for individual liberty. (Rothbard 1982, p. xlvi)

However, this is still not the bedrock: the case for individual liberty is in the last resort established by a transcendental argument that shows that individual liberty cannot be denied without committing a performative contradiction. Indeed, following Rothbard's normative turn, Hans-Hermann Hoppe (1989, chapter 7) provides the link with the transcendental argument for freedom and freedom rights (see also most recently Hoppe 2023). *Hence, ultimately, the intellectual currents briefly rehearsed in this section and the preceding one all come together in this normative view of persons and their inalienable rights.*

In addition to the knowledge acquired by the empirical sciences, there is thus a knowledge that relates to thought and action itself. This knowledge reveals the connection between freedom, reason and normativity. Based on the fact that in the case of thoughts and actions—and only in the case of these—it makes sense to ask for reasons, this knowledge demonstrates freedom in thought and action and the normativity associated with it through a transcendental argument. However, this is merely a knowledge of freedom in forming judgements and intentions. It is not substantive, positive knowledge about goals of action, values, the common good or the meaning of life. Corresponding knowledge claims satisfy neither the requirements of empirical science nor those of transcendental arguments.

This leads us to the following conclusion: recognising someone as a person is to adopt a normative attitude towards them—to recognise their utterances as thoughts and their behaviour as actions for which reasons can be demanded and which are thus subject to a judgement as correct or incorrect. However, action is subject to reasons without there being any knowledge in terms of content about what one should do. That is why reason is always only a means and never as such a purpose in life. There are no goals in life or a meaning of life that reason could positively provide. Normative knowledge only concerns the universalisation of freedom: to recognise someone as a person is to grant that being freedom in thought and action. And this means that whatever goals a person sets for her- or himself, these goals and the ways to achieve them must be compatible, in the sense of a moral obligation, with recognising the freedom of every other person to set goals for her or his actions as well. This, then, is the basis of the rule of law.

It is thus evident that denouncing this concept of freedom as promoting egoism is a complete distortion. Freedom, as the Enlightenment, classical liberalism and libertarianism conceive it, implies the moral *obligation* for every person to unconditionally recognise the freedom rights of every other person—quoting again Kant: to always treat every person as an end in itself and never as a mere means to an end. This, then, is the tightrope that the use of reason requires us to walk: there is objective knowledge also in the humanities. But this is only a formal knowledge of the freedom that each person enjoys in thinking and acting and thus only a normative knowledge of the moral obligation to respect this freedom. It cannot be a knowledge of a normative content of goals, values or meaning for the life of a person or a common good for a society.

Consequently, political scientism is morally reprehensible, because it contradicts the moral obligation to always treat every person as an end in itself and never as a mere means to achieve an end. In any regime guided by political scientism, some persons are treated as mere means to achieve the end set by "The Science"—such as in the corona regime, the persons whose physical or mental health is destroyed by the lockdowns under the pretext of "public health protection", or the children and adolescents who suffer from health damage caused by the vaccinations that are imposed on them under the pretext of "solidarity". Political scientism therefore turns the relationship between science and human freedom upside down. Human freedom is a presupposition for science. Science in the sense of modern natural science cannot grasp its prerequisites in human thought and action.

Restoring science hence requires elaborating on the distinction between on the one hand natural science that is empirical, a posteriori and about matters of fact and on the other hand a science of human thought and action that is based on transcendental arguments, a priori and normative in that it presupposes the normative stance in which humans are recognised as persons and, consequently, the moral obligation to treat them as persons. The intellectual tools how to do so are all here: Descartes as well as Hayek and Popper establish why the very aim of objectivity of the natural sciences make them inapplicable when it comes to human thought and action. Kant provides the transcendental argument that establishes human freedom and freedom rights. The linguistic turn in philosophy in the twentieth century enables us to show how what Kant considers to be a transcendental subject is implemented in the social, normative relations that are constitutive of human actions, as elaborated on by the Pittsburgh school of Sellars (1956, 1962), Brandom (1994) and McDowell (1994, 1995) among others. In the Austrian School of economics, Mises (1949) works out how the social science of economics is an a priori science, following from human action. The normative turn in Rothbard (1982) brings in normativity and with Hoppe (1989, chapter 7) also the link with transcendental arguments. Hence, all these intellectual currents come together in a normative, a priori science of human freedom in thought and action and its consequences. Thus, the ingredients for the intellectual cure against scientism and its political use are all in our minds and hands. We only have to communicate them properly to the public.

However, even if this is sufficient to set up a positive and constructive alternative to scientism, it is not sufficient to dismantle political scientism. There is a (pseudo-)scientific ingredient to political scientism, but there is also a *sui generis* political one. It is not the case that scientists on their own usurp politics. Quite to the contrary, they are called in by politicians to legitimise their extension of power in regulating ever more aspects of social and even private life (as in the example discussed at the end of Chap. 2, Sect. 2.7 with the German chancellor Angela Merkel making use of the National Academy of Sciences Leopoldina to legitimise comprehensive social control in the corona regime). As there is a mechanism that leads from science to scientism, so there also is a mechanism that leads from the rule of law as implemented in the modern constitutional state to the welfare state that seeks to extend its power in protecting people from ever more risks (up to and including risks stemming from the spread of viruses, climate change or the expression of views that may hurt the feelings of someone). In doing so, the welfare state then promotes the staging of an alleged science that identifies the risks in question and shows the way out through a comprehensive regulation of social and even private life, thus ending up in welfare totalitarianism. Hence, in parallel to restoring science, we also have to restore the rule of law.

CHAPTER 4

The Rule of Law Versus Welfare Totalitarianism

4.1 From the Presumption of Liberty to the Justification of Coercion

Science is one hallmark of modernity. The rule of law and its implementation in constitutional states, leading to liberal democracies, is the other one. Modernity witnessed not only the development of physics, chemistry and biology to levels never seen before but also the development of liberalism and the rule of law. However, as the success of science led to scientism, so the success of liberalism led to an overexpansion of states in the form of welfare states. Welfare states have increasingly broken the liberal rule of law and undermined the basis of liberal democracies: the more extensive the welfare network is, the less strong the rule of law is. The danger of welfare states is that they incur in totalitarian tendencies that eventually lead to *welfare totalitarianism*, imposing a certain view of the common good and invoking "The Science" for its legitimation. In that sense, overextended welfare states are to liberal democracies and the rule of law what scientism is to modern science. They coincide in political scientism becoming a sort of state religion that legitimises an all-comprehensive control of social and even private life.

In order to understand how this is possible, we retrace in this chapter the path from human freedom to freedom rights and natural law and from there to the rule of law. We elaborate on the argument to implement the

M. Esfeld, C. Lopez, *Restoring Science and the Rule of Law*, Palgrave Studies in Austrian Economics,
https://doi.org/10.1007/978-3-031-71186-2_4

rule of law by constitutional states that develop into liberal democracies. But we also point out the dilemma that lies in the justification of these states and that is already evident in Thomas Hobbes' masterpiece *Leviathan* (1651). We then show how, as a result of this dilemma, constitutional states tend to turn into welfare states that are prone to welfare totalitarianism. Against this background, we finally set out at the end of this chapter our explanation of why the rule of law and the mechanisms for controlling the exercise of power could so easily fade away in the corona, the climate and the wokeness regimes. The explanation is what we call "the illusion of republicanism". Once we have understood that illusion, we can free ourselves from it and restore the rule of law and thereby also the framework in which science can blossom. We show how to do so in Chap. 5 by making use of the intellectual resources of Austrian libertarianism.

As explained in Chap. 2, Sect. 2.1, scientism and its political use derive from the Platonic tradition of knowledge claims about the common good by an elite of experts, such as intellectuals, priests or scientists. However, neither modern natural science nor the normative knowledge obtained by reflecting on human thought and action (transcendental argument) could deliver such knowledge. Consequently, there cannot be a justification stemming from science for suspending freedom rights. Both modern science and the modern reflection on persons as subjects of thought and action vindicate the Socratic-Aristotelian tradition.

Christianity stands between antiquity and modernity. Historically, the idea of universal human rights goes back to the Christian doctrine of the creation of human beings in the image of God. Because God created humans as beings who think and act and who are thus free in their thinking and acting, it is morally reprehensible to impose something on other people against their will that one estimates having identified as the common good, even if the corresponding knowledge claims were true. Platonic rule by an elite of experts following "The Science" is therefore incompatible with the Christian religion—even if there really were experts who have knowledge of the common good. Political scientism hence is also deeply anti-Christian. It jettisons everything that Christian religion has brought to human civilisation.

In modern times, the Christian idea of universal human rights is secularised: their justification no longer depends on the Christian religious doctrine. The justification by a transcendental argument elaborated on in Chap. 3. Sect. 3.4 refers only to the use of reason that is common to all human beings, irrespective of their religion, culture, ethnicity, gender etc.

We thus get to the *presumption of liberty* that is characteristic of the modern liberal tradition. Gerald Gaus expresses this presumption in this way:

> The liberal tradition in political philosophy maintains that each person is free to do as he wishes until some justification is offered for limiting his liberty … we necessarily claim liberty to act as we see fit unless reason can be provided for interfering. (Gaus 2003, p. 207)

The presumption of liberty covers politically the whole spectrum from left to right (e.g. Feinberg 1973; Rawls 2001; Nozick 1974). It is not only a normative principle for political philosophy but also dialectically powerful. It puts the burden of proof on those who advocate coercion. Coercion is what calls for justification, not granting freedom and freedom rights. This is the basis of all modern political philosophy.

Coercion is the use of violence—physical force—against a person such as locking them up, hurting their body or even killing them (see Huemer 2013, pp. 8–9). A person is also coerced to provide goods or services if she is threatened with the use of violence against her in case she refuses to do so. Coercion thus consists in intentionally depriving a person from making use of her time and body in a self-determined way. Applying coercion thereby entails not to respect a person as an end in itself, but to treat her as a mere means to achieve an end. This does not entail that coercion is never justified. Individuals incur in what philosophers of law call right-forfeiting actions, that is, actions through which an individual loses a right, as, for instance, in case of a robbery, murder or unprovoked violence. Consequently, *coercion is legitimate and justified if and only if the person against whom coercion is applied first did something—or set out to do something—through which she encroaches upon the way other persons fathom their life*, that is, treats other persons not as an end in themselves, but as a mere means to her ends. Coercion then is legitimate as defence against such an act (such as self-defence against an assault) or as retaliation against a person for having done such an act.

In common sense morality, coercion is considered morally wrong without counterbalancing reasons that justify why one is entitled to use force or to threaten to use force against another person. Think of the following pedestrian example: my neighbours fall sick from an unknown disease. Afraid of me and my family becoming infected, I put up a barricade in front of their door and threaten to shoot them if they dare to go outside. It is clear that I do not have any right to do that. I may avoid direct contact with my neighbours or take other protective measures for myself and my family such as strengthening our immune systems. But I must not lock my neighbours up. Hence, common sense morality works on the basis of the presumption of liberty and the need to justify coercion.

To put this crucial point differently, from the normativity to which human freedom in thought and action is tied follows the categorical moral obligation to always treat every person as an end in itself and never as a mere means to an end—and be that end a common good. From this moral obligation then follows that coercion has to be justified; for coercion means treating a person not as an end in itself. It follows in particular that claims to knowledge of the common good, even if these claims were true, could never justify coercion to achieve the common good; for this always implies treating some people as mere means to achieve the—alleged—common good. If one thinks one knows the common good, one has to convince people to act accordingly out of their own free will. One must never coerce them to do so. From the mentioned moral obligation follows that applying coercion against a person is justified if and only if that person violates that obligation through some of her actions.

This argument has far-reaching consequences, which are by no means respected across the whole political spectrum that defines the modern liberal tradition, although these consequences follow from the presumption of liberty. The failure to respect these consequences is at the roots of the destruction of science and the collapse of the rule of law that we undergo in the corona, climate and wokeness regimes, as we shall explain in this chapter.

In the first place, we have to distinguish between two notions of freedom or liberty that can be dubbed "positive" and "negative" freedom, following an influential essay by Isiah Berlin (1969). Freedom as a positive concept signifies self-determination in thought and action. Thus, *positive freedom* is "the answer to the question 'What, or who, is the source of control or interference that can determine someone to do, or be, this rather than that?'" (Berlin 1969, p. 121). Freedom as a negative concept means non-interference. Thus, *negative freedom* is "the answer to the question 'What is the area within the subject—a person or a group of persons—is or should be left to do or be what he is able to do or be, without interference by other persons?'" (Berlin 1969, p. 121).

The distinction between negative and positive freedom can also be found in Kant. Negative freedom is independence:

> That *independence*, however, is freedom in the *negative* sense, whereas this *lawgiving of its own* on the pure and, as such, practical reason is freedom in the *positive* sense. (*Critique of Practical Reason* 1788, book 1, chapter 1, § 8, theorem 4; translation adopted from Kant 1996, p. 166)

The crucial point then is that although freedom is a positive concept in the first place, signifying autonomy or self-determination, no positive *rights* follow from this conception of freedom. The freedom rights that are entailed by the positive notion of freedom—the trio of freedom, reason and normativity as explained in Chap. 3, Sect. 3.3—are only negative rights of defence against unwanted interference in the way in which a person self-determines her life. The moral obligation to always treat every person as an end in itself and never as a mere means to achieve an end is the obligation to refrain from interfering with the way in which a person self-determines her life. This obligation thus establishes negative freedom rights to defence against unsolicited interference, but it cannot establish any positive freedom rights.

One may recognise a right to freedom in the sense of the definition of a right according to which having a right to something amounts to enjoying some moral status (see Thomson 1990, p. 38): there is some moral significance in having such a right. The moral significance of rights derives from their correlation with obligations. This means that the rights that an individual *A* possesses constrain or limit the actions that other individuals may do against *A*. Paraphrasing H. L. A. Hart's way to put it, to have a right is to have a moral justification for limiting other persons' behaviour towards one (Hart 1955, p. 183). This expresses the well-known right-duty correlation (see Hart 1955; Beran 1987; Thomson 1990). In general, the right-holder is entitled to have the duty-bearer *do* or *refrain* from some act. That is, to have a right constrains the behaviour of others, since there are things they have to do or refrain to do in virtue of one's having a right. In this way, to have a right to freedom is to enjoy a moral status that imposes restrictions upon others about what they can do against a person with respect to her actions, decisions and thoughts. In this sense, a right to freedom offers normative protection for one's actions, decisions and thoughts.

However, the right to freedom consists only in the right to defence against unsolicited interference with the manner one makes up one's mind in one's actions, decisions and thoughts. This right imposes upon others only the moral obligation to refrain from encroaching upon the way in which a person fathoms her life in her actions, decisions and thoughts. The right to freedom hence is only a negative right. Positive rights, by contrast, are rights to certain entitlements. If a person has a right to something, she is entitled to being provided with the thing in question. In virtue of the right-duty correlation, others thus have the obligation to act such that the entitlement in question is fulfilled. Hence, these obligations are not merely duties of non-interference, but they may go well beyond.

That notwithstanding, there obviously are positive rights. Entitlement rights undoubtedly exist. But they cannot be universal. They come into being through voluntary contracts among persons. For instance, if a person agrees to do some work that is assigned to her by another person and carries out that work, she thereby acquires the entitlement to be compensated by that other person; accordingly, that other person has the moral obligation to compensate her by providing some specified goods or services (unless otherwise agreed by the two parties). Positive or entitlement rights in this sense are conditional rights: their origin is some voluntary agreement among persons.

However, when positive rights are construed as basic and universal, holding unconditionally, a contradiction arises: unconditional rights to certain entitlements contradict the moral obligation to always treat each person as an end in itself and never as a mere means to achieve an end. The reason is that to any entitlement of a person to something corresponds the obligation of other persons to provide the good or service in question. Thus, if a right to certain benefits—such as to being provided with clean water, food, clothes and housing—obtained in virtue of universal positive rights, then there would also have to be an inherent obligation for every other person to fulfil the entitlement rights of these persons. People would be categorically obliged to spend time of their life and work to do certain things.

To make this example concrete, suppose that everybody were entitled to an unconditional basic income that allows them to purchase drink, food, clothes, housing and the like as a material basis for leading a self-determined life. Fulfilling this entitlement would imply that some people have to be forced—through the employment of coercion, if need be—to spend time of their life to do the work such that everybody receives the basic income. As an unconditional obligation that obtains in virtue of universal positive rights, this would clearly violate their negative freedom against unwanted interference with how they spend their time and what work they do. It would amount to unconditionally enforced labour.

Coercion would be exercised against certain people without these people having done any actions through which they treat other persons merely as a mere means to achieve their ends. The bottom line hence is that a right cannot be both positive and universal, holding unconditionally.

Furthermore, even if positive rights in the guise of entitlement rights obtained unconditionally, they could yet not be fulfilled unconditionally. Whether every person can be provided with a basic income that allows them to purchase clean water, food, clothes and housing depends on contingent circumstances that may obtain or not. Obligations that hold in virtue of the normative nature of human freedom, by contrast, have to be such that they can be fulfilled unconditionally. Refraining from encroaching upon the course of life of others is a general rule than can always be respected and under all circumstances, whereas providing some particular goods and services to others is always subject to contingent circumstances of availability.

The negative freedom rights that are entailed by the trio of freedom, reason and normativity are ***natural rights***, because they are the rights that follow from the—normative—human nature as thinking and acting beings as explained in Chap. 3, Sect. 3.3. These negative freedom rights are the only natural rights. Accordingly, the moral obligation to always treat every person as an end in itself and never as a mere means to achieve an end is the essence of ***natural law***; this is the law that is derived from—normative—human nature as consisting in being endowed with reason and thereby with freedom.

In Christianity, the justification of rights emanates from God and depends on the equality of the souls in front of their creator. The intrinsic dignity of human beings without distinctions as well as their natural freedom are constantly emphasised in the biblical tradition (see Villa-Vicencio 2000). In modernity, the idea of universal rights is secularised in the form of ***natural*** rights and ***natural*** law, which are accessible through reason (see Tierney 1982 for the medieval-modernity transition). Many of the Christian distinctive elements such as universality and equality are retained. The same natural rights—and obligations—apply to all human beings independently of their ethnicity, religion, social status, gender etc. The modern emphasis is on reason as a way to know and justify these rights. The justification by a transcendental argument elaborated on in Chap. 3, Sect. 3.4 refers to the use of reason in thought and action, which unites all human beings independently of their ethnicity, religion, social status, gender etc.

Natural law and the obligations correlated with natural rights demarcate only a minimal morality—the one that can be established by the use of universal reason alone through a transcendental argument and that, accordingly, holds a priori and unconditionally. It expresses the common essence of all moral systems including, in particular, the Christian ten commandments. However, obviously, particular ethical systems typically contain moral obligations that go far beyond the obligation to always refrain from unsolicited interference with the lives of other people. They notably establish moral obligations to do something on behalf of others. Such ethical systems may well be justified on religious or humanitarian grounds. But this is not a matter of natural law.

Natural law is such that coercion is justified to enforce it. That is exactly what the idea of the *rule of law* expresses. The rule of law signifies enforcing the respect of natural law through coercion within a legal system. By contrast, moral obligations to do something on behalf of others cannot be realised through coercion. To the extent that the demands of morality are met, they have to be fulfilled voluntarily and are therefore virtuous actions.

What constitutes an unsolicited interference with the course of life of other persons such that these persons have a natural freedom right of defence against such an interference? Obviously, any attack against the life and limbs of a person—the application of physical force or the threat to do so—constitutes such an interference as well as any attack against her property (unless, of course, the attack is a reaction to a previous wrongdoing of the person under attack). The crucial issue then is to specify how a person acquires property rights to things under natural law. If there were no natural property rights and all property rights depended on positive law, the fruits of the labour of a person would not be her property independently of a law-making procedure. However, this would be paradoxical: one of the central arguments in favour of the existence of the state is that it is meant to protect property rights. But it would be conceptually unclear which property rights it should protect if they depended on the law-making procedure—that is, if individuals did not already enjoy property rights independently of the constitution of the state. Thus, most arguments in favour of the existence of the state (or, in particular, of positive law) already assume the existence of natural property rights.

How property rights are acquired is certainly a difficult topic. It is clear that self-ownership, namely property rights on one's own body are primary. A person's rights of defence against unsolicited interference in the way she sets goals for her life and realises these goals entail property rights on how she employs her bodily capacities. Hence, the way in which a person fathoms her life also includes how she uses her capacities to produce external goods (such as tools, land etc.). This takes property rights beyond self-ownership: it includes the possibility to acquire property rights on external things. According to John Locke's famous theory of the acquisition of property in his *Two Treatises of Government* (1689, second treatise, chapter 5), one acquires a property right in something by mixing labour in the thing in question (labour theory of property acquisition or principle of first appropriation).

Locke's theory aims to explain how things that are external to us can be legitimately owned. More precisely, the theory establishes that there is a right to claim ownership of previously unowned things on the condition that one has mixed one's labour with these things—that is, uses these things as a means to achieve a goal that one sets oneself for one's life. Thus, for instance, if an artist uses unowned clay to build a statue, the clay transformed into a statue becomes her property. If people use previously unowned land to grow corn or vegetables on it or to extract commodities from it, they thereby acquire a property right in the products and the land, at least insofar as and as long as they employ the land to produce them. Thus, merely discovering a previously unchartered territory and building a fence around it is not sufficient to acquire a property right in it; such a right is acquired only to the extent that one invests labour in the land. This theory is pragmatically useful since it allows us to avoid cases like the so-called tragedy of the commons (Hardin 1968), since property rights can be legitimately claimed. It also rules out cases of illegitimate appropriation as when an unchartered territory is fenced.

Things legitimately acquired through the first appropriation can subsequently be handed over voluntarily to others. Voluntary contracts, for instance when one sells or buys something, transfer property rights. Inheritance is also a case of legitimate transfer of property rights. There are of course many grey zones. For instance, are there collective property rights? Can a community own a portion of land, or can workers own a factory collectively? Or do they own each a share of the land or of the factory such that each is free to transfer her or his share to others through voluntary contracts? In general, insofar as the agreements are voluntary and there exists a clear demarcation of what one is entitled to when claiming property rights, many forms of agreements are conceivable.

Against this background, it is possible to operationalise all natural freedom rights as property rights. Every person has an original property right in her life and the disposal of her body (self-ownership) plus derivative property rights in external things acquired through labour or transference. If a person possesses these rights, then she can make use of them as she sees fit, that is, she is free to dispose of her property as she pleases. She thus has the right to defence against unsolicited interference *because* her property is *her* property, and an unsolicited interference violates that property right. Hence, freedom rights are a way to exert property rights.

The operationalisation in terms of property rights avoids possible conflicts within natural law. Thus, for instance, a person has the right to use her voice as she thinks fit, that is, the right to free speech. However, if she is on premises owned by other persons, these other persons have the right to limit speech, to forbid the expression of views they consider offensive on their premises, to allocate slots to speakers etc. in virtue of their right of defence against encroachment upon the use of their property. In the vein of Austrian libertarianism, Murray Rothbard (1982, part II, in particular chapter 15) sets out a detailed account of natural rights in terms of property rights (see also Hoppe 1989, in particular chapter 7, and Block 2019). In short, the negative freedom rights of defence against unwanted interference with the way in which one fathoms one's life need to be operationalised as property rights in order to be applicable. There is no enforcement of natural rights and the rule of law without the recognition of natural property rights.

4.2 From Natural Law to the Rule of Law

Immanuel Kant says in his lecture *Natural Law* (1784):

> Law is the restriction of freedom, according to which it can exist with every other person's freedom according to a general rule. … But if everyone were free without law, nothing more terrible could be conceived. For each would do with the other as he wished, and so no one would be free. … Law is therefore based on the restriction of freedom. … In the case of law, happiness does not come into consideration at all; for everyone can seek to attain it as he wishes. (Kant 1979, pp. 1320–21, our own translation)

As the title of the lecture already indicates, Kant means by "law" in this context "natural law". Natural law is natural in the sense that it derives from human nature, in its normative character as set out in Chap. 3, Sect. 3.3 and further developed in the preceding section: freedom is the nature of human beings, and freedom is tied to reason and normativity. From this normative nature of humans follow natural freedom rights and with them natural law as a normative conclusion.

However, natural law consists only in the universalisation of freedom, namely in the commandment to make use of one's own freedom to shape one's life only to the extent that this use is compatible with the freedom of every other person to shape their life. That is, freedom is exercised in a framework of mutually constraining duties. With the help of the transcendental argument and the aforementioned connection between freedom and normativity, natural law can be clearly defined: it enshrines the right of defence against undesired interference in the way in which a person fathoms her life.

Indeed, the general rule that makes the freedom of every person compatible with the freedom of every other person is the rule to exercise one's self-determination only in such a way that one does not encroach upon the self-determination of other persons. *Natural law consists only in negative rights*, namely rights of defence against unwanted external interference. Consequently, the rule of law implements only negative freedom. That is, it is the framework that protects the diverse ways in which positive freedom can be exercised. The rule of law cannot implement positive freedom insofar as granting positive rights unconditionally would constitute an encroachment upon the self-determination of persons as explained in the preceding section, giving rise to a contradiction. Positive freedom is what characterises human nature and is the basis from which natural law derives. However, the rule of law cannot impose a uniform way to exercise it, but only a uniform rule to protect its exercise. As Kant says, "happiness does not come into consideration at all" as far as natural law is concerned. There are no obligations as far as natural rights and natural law are concerned that refer to the fulfilment of material conditions for happiness.

Hence, the very notion of natural law—and the rule of law as being based on natural law—is incompatible with legal positivism. For something to be law in the sense of the rule of law, it is neither necessary nor sufficient that it was adopted in accordance with a certain formal procedure of law-making in a political community. If such a procedure were sufficient, then more or less every dictatorship would count as implementing the rule of law. Even the totalitarian socialist regimes of the twentieth century of the communist and the national socialist type would then fall under the rule of law. However, far from being void of content, natural law is a sharp normative notion: existing positive laws can be criticised in its light, namely as being incompatible with natural law. Positive laws may well fall short of meeting the standard of natural law and thereby the rule of law. Consequently, the notion of the rule of law serves as a court of appeal in the face of unjust positive laws.

This normative criterion also applies when the formal procedure of law making includes a popular vote, as in a democracy, or even a direct democracy in which every positive law has to be adopted by a popular vote. Also, positive laws approved by the majority of the people on a territory who are subject to the law in question can be incompatible with the rule of law. This is obvious when turning to human rights: a majority vote that deprives a minority of—some—of their human rights would clearly be illegitimate in the light of the rule of law. This can be so even if one holds that a popular vote is the best available procedure to settle whether or not a given law proposal is compatible with human rights and whether or not it meets the standard of the rule of law, instead of delegating the decision on that issue to experts in the guise of judges. That is the reason why Switzerland, for example, does not have a constitutional court. One must distinguish between the rule of law as normative notion and the practical question of what is the best procedure to implement it, given that all conceivable procedures are fallible.

All this implies that, as far as normativity is concerned, the rule of law is the bedrock of modern society. The obligation to comply with the rule of law holds unconditionally. If the rule of law holds unconditionally, then what the law in the sense of natural law demands must be such that it can always be fulfilled. Coming back to Kant's definition of law, if natural law is "the restriction of freedom, according to which it can exist with every other person's freedom according to a general rule", then it must always be possible for a person to act in compliance with that general rule. This, in turn, implies that natural law is a matter of cognition instead of stipulation and that this cognition is available to everybody by using their own reason instead of requiring expert knowledge.

As objectivity in natural science—the point of view from nowhere—so also the rule of law is a regulative idea. Again, strictly speaking, the very idea of the rule of law is contradictory: law cannot rule, only humans can. The rule of law, implementing natural law, is a regulative idea that is intended to guide any form of human rule. It serves as a standard on the basis of which existing forms of rule can be assessed. In the same way as only a true statement can be knowledge and the truth of the statement does not depend on the person who utters it and their social community, so only something that is compatible with natural law can be law. Like science, also law is about knowledge and not about conventions.

Whether one drives on the left or on the right in traffic is a matter of positing (convention) instead of knowledge (cognition). Once such a convention is in force, all those who want to use the traffic infrastructure are obliged to respect this convention. However, law has a different origin. The question of where the boundary lies beyond which a person makes use of her freedom in such a way that she encroaches upon the freedom rights of other persons is a matter of knowledge instead of stipulation. Whether robbery is wrong is not matter of convention or of the community's legal codes; it is a matter of knowledge of what one is morally obliged to do in treating persons as ends in themselves and hence of what respects the natural law and what does not. Law thus has the same origin as scientific knowledge, namely, to find out what is correct. The idea of making law instead of finding out what the law—namely the natural law—demands in a concrete situation is misguided. It is as wrong-headed as the idea that we make the laws of nature such as the law of gravity (see the remarks on intellectual postmodernism in Chap. 2, Sect. 2.6).

There hence is a parallel between the laws of nature that science discovers and natural law in the normative sphere of human actions. That notwithstanding, laws of nature such as the law of gravity determine the motion of matter given initial conditions—provided that they are deterministic; otherwise, they determine objective probabilities for the motion of matter.[1] By contrast, natural law does not determine human social interactions: it does not prescribe a content for what we should do. It only sets a frame that is to be respected in anything that we do. This, again, is the tightrope that the use of reason requires us to walk (see the end of Chap. 3, Sect. 3.4): law is in any case a matter of discovery. In the case of the natural sciences, this discovery enables technical engineering. By contrast, social engineering when it comes to society is contrary to law and destroys the rule of law; for the law in the case of social interactions is the universalisation of freedom and freedom rights, consisting in the obligation to always respect every person as an end in itself and never treat a person as a mere means. There is no objective knowledge of goals, neither in natural science (because the motion of matter does not have a goal), nor in the

[1] See Esfeld (2020, chapter 2.4) for a detailed argument why the recognition of laws of nature, even if they are deterministic, does not entail a conflict with human free will and its effects on human behaviour.

humanities (because the knowledge concerns human freedom). That, again, is why political scientism goes astray.

As explained at the end of Chap. 2, Sect. 2.1, in the case of knowledge in the sense of scientific knowledge, no authority is required to determine what the corpus of scientific knowledge is. There is no rule required to establish anything as scientific knowledge and enforce it under threat and, if necessary, use of coercion. When it comes to law, by contrast, the situation is different: a procedure is necessary to bring about a decision in the case of disputes about what natural law demands in a given situation. This is a matter of knowledge, as in science: claims to knowledge in science never violate the freedom rights of persons (unless experiments are done to test such claims that affect other persons against their will). Actions, by contrast, are subject to the duty of always respecting every person as an end in itself. If persons violate this duty in some of their actions, the use of coercion against these persons is justified.

The problem is then epistemic: how do we know in a concrete situation that an individual or a group of individuals has violated other persons' rights and deserves punishment? To resort to violence to settle the issue would be a far greater evil than setting up a legal procedure to do so, although any such procedure is poised to be fallible. That is why societies have generated various mechanisms to settle these disputes by invoking a third, non-involved party. The judiciary system that brings about a decision to settle a conflict by a judge is such a mechanism. In that case, the in-conflict parties submit their case to a third party in the guise of a judge that, then, is entitled to enforce its decision through coercion. That is, the use of physical force is legitimate only if approved by a law court that heard and judged the case. Natural law therefore requires establishing legal mechanisms to resolve conflicts that include clear and known procedures that, although fallible, establish and enforce something as corresponding to the law in a given situation. That is what the term "rule of law" is intended to express.

That notwithstanding, as such, natural law implies nothing in terms of *specific* procedures and institutions for gaining and enforcing knowledge of law. Natural law can be unambiguously determined by the use of reason as outlined in Chap. 3, Sect. 3.3. Although knowing what the natural law demands in a particular case is an epistemic issue, the enforcement of the natural law is a practical problem. There is no privileged knowledge of law embodied in particular institutions or persons. As far as the justification of epistemic claims is concerned, the same anarchy holds in the case of law as in the case of scientific knowledge: there is no other instance of justification than evidence and argument, which all participants can assess by using their reason. Everything else is a practical matter, which in the case of law—unlike scientific knowledge—requires regulation. Thus, judges do not have a privileged expert knowledge that is not accessible to the general public. They are trained and qualified to apply certain procedural rules concerning the fair hearing of cases in which one party accuses another party of violating their rights through some of their actions.

A legal order worthy of the name accordingly is the attempt to cast the knowledge of law in the sense of natural law into the form of institutions that fix and enforce this law. There are three arguments for understanding natural law in the aforementioned sense as the basis of the legal order and thus at the same time as that which sets the limits of what is to be regulated by a legal order:

- *The argument of epistemic modesty*: There is no knowledge of a common good that would entitle those who claim such knowledge to rule over others and prescribe how they should live their lives. People can voluntarily join religious or other communities and commit themselves to following the instructions of leaders or experts on how to shape their lives. But there is no justification for those who claim to have expert knowledge to impose what they consider to be the common good on others through coercion.
- *The normative argument of freedom rights*: Freedom rights as natural rights follow from the normative character of freedom in thought and action that is at the core of human nature. These are rights of defence against unsolicited interference in one's own way of life, but not rights to the fulfilment of certain claims (no entitlement rights). Coercion is justified if and only if a person violates the freedom rights of other persons through some of her actions—that is, fails to respect other persons as an end in themselves and treats them merely as a means to some of her ends. Accordingly, a legal order based on natural law must not coerce anybody to act such that others attain what they consider to be their happiness, as Kant puts it in the quotation at the beginning of this section. Again, particular ethical systems may impose such obligations, and they may well be justified on humanitarian or religions grounds. But the rule of law cannot do so.
- *The argument of consequences*: Respect of the freedom rights of everyone is the most successful path to technological and economic

progress and thereby better living conditions that benefit all segments of the population. According to Mises (1949), this follows a priori from the theory of action, as explained in Chap. 3, Sect. 3.4. Empirically, the following statistical correlation is well established, which one tries to measure, for example, in a freedom index (see Vasquez et al. 2023): the closer a legal order comes to the ideal of the rule of law, the higher the standard of living is in the population as a whole. This correlation is an important element for the acceptance of the rule of law: its respect in the population is based not only on regarding freedom from unwanted external interference as a value in itself but also on the fact that this legal order paves the way to enormous technological and economic progress, which improves the standard of living and quality of life for all segments of the population. One of the main reasons is that a firmly established rule of law delivers predictability and long-term order, which economic agents interpret as less risks and lower costs.

4.3 The Constitutional State and Its Dilemmas

Modern political philosophy, mainly in its liberal strain, has been built on the presumption of liberty, as explained in the previous section. The so-called state of nature is the par excellence analytical device on the basis of which a coercive legal order is to be justified. In John Locke's words:

> To understand political power aright, and derive it from its original, we must consider what estate all men are naturally in, and that is, a state of perfect freedom to order their actions, and dispose of their possessions and persons as they think fit, within the bounds of the law of Nature, without asking leave or depending upon the will of any other man. (*Two Treatises of Government*, 1690, second treatise, chapter II, § 4; quoted from the edition Locke 1823)

The state of nature has, however, given place to many misinterpretations. In the liberal tradition, it is not a state of isolated individuals. Humans are social beings. What distinguishes humans from other social animals is the trio of freedom, reason and normativity that comes with thought and action, as explained in Chap. 3, Sects. 3.3 and 3.4. This normativity and social feature of the state of nature is expressed by Locke in terms of "the bounds of the law of nature".

More precisely, the specific feature that distinguishes the social order of humans from the social order of other animals is that this order is a *normative* order, in which individuals recognise each other as moral beings, having rights and obligations. The common framework that comprehends these moral relations is the implementation of the rule of law. Again, the use of universal reason that unites all human beings is the guide to set up such an order. From this it follows, again, that the employment of reason is a means to limit the use of power: claims to power have to meet the standard set by reason; they have to comply with the regulative idea of the rule of law within which all individuals relate to each other by moral-normative, social relations.

The origin of the liberal tradition and constitutional states is complex. It arguably took roots in the late medieval age in the proto-constitutionalism and civic humanism of the Italian medieval cities (Tierney 1982; Hankins 2000). Our focus here is on the conceptual fundamentals and salient features of political liberalism in modernity. One of its central elements is the above-mentioned secularisation of the idea of *rights*—that is, the secularisation of the idea that the same rights apply to all.

A second element following from this one is the idea of a *universal* protection and *impersonal* guarantee of such rights. This has led to the attempt to implement the rule of law through a number of formal and procedural principles that extend and enforce rights universally, protecting persons from external and arbitrary aggression. In this way, modernity and the entire liberal tradition view the rule of law as an ideal that imposes impersonal, public, clear and universal norms that protect everyone. Indeed, the presumption of liberty along with the respect of property rights are central tenets in the ideal of the rule of law since the early modernity. From the nineteenth century onwards, the practical extension of the rule of law and its institutionalisation in constitutional states led to a universal enforcement of basic rights. For instance, the abolition of slavery, which originated in England, was an essential achievement in this line, as was later the banning of all sorts of publicly displayed racism in the second half of the twentieth century.

The third element follows from what has been said so far: political power calls for a justification in a secular way. John Locke's *Second treatise of government* published in 1690 is one of the milestones in the development of the liberal tradition and the route to constitutional states. Locke's intentions are clear from the very beginning: he attempts to overthrow Robert Filmer's theological justification of the monarchic state, offering instead a secular justification of constitutional republics. Locke's work was published almost 40 years later than the *Leviathan* (1651) by Thomas Hobbes, where the nature of political power is brilliantly exposed. Hobbes' work is also one of the earliest and most influential attempts to legitimise the state in a secular way.

In the previous section, we mentioned that one of the pillars of modern political philosophy (and the core of the liberal tradition) is the presumption of liberty. This entails that coercion is what calls for justification, not liberty. Indeed, the hallmark of the state is that its institutions settle when the use of force (physical violence) is legitimate and when it is not: according to Max Weber (1919, § 4, and 1922, chapter I, § 17), the state is best defined by exercising the monopoly of allowing and forbidding the use of force on a certain territory. Robert Nozick expresses this characterisation of the state in these terms:

> A state claims a monopoly on deciding who may use force and when; it says that only it may decide who may use force and under what conditions; it reserves to itself the sole right to pass on the legitimacy and permissibility of any use of force within its boundaries; furthermore it claims the right to punish all those who violate its claimed monopoly. (Nozick 1974, p. 23)

This means that a state arises when a group of persons reserves to itself the right to settle who is authorised to enforce rights, to exercise punishment and to deliver justice within a territory. Even if normativity distinguishes the social order of humans from the social order of other animals, we still need a justification for the transition from natural law as a *moral* obligation to the implementation of the rule of law in the guise of a *state* whose organs claim the *monopoly of law making* in the sense of casting natural law into precise positive laws that are imposed on the people living on a certain territory (legislative), *law enforcement* (executive) and *judiciary*.

In the Lockean state of nature, one of the relevant features is the natural symmetry that exists among individuals to interpret and to enforce natural law. Besides that the state-of-nature device is meant to play an explanatory role in political philosophy, it is clear that it presumes that all human beings are politically equal. In a state that claims the monopoly of the use of force, by contrast, some people enjoy political authority by holding offices and others owe political obligations to them (see Simmons 1979). Hence, there is the need to justify the political asymmetry that comes with the state.

The most widespread argument to legitimise the state is that it is the best—or even the only—way to protect the natural rights of the individuals and their social communities such as families and to implement the rule of law. More precisely, what is at issue here is the modern constitutional state as the political organisation of the modern society that is no longer organised along the lines of kinship and small communities based on personal acquaintance. Social interactions bring people together with whom one is not acquainted. Consequently, a problem of trust arises: Are they reliable partners for interaction? Will they respect contracts? The modern constitutional state is the solution to this problem of mutual trust and large-scale coordination. In short, by enforcing natural law, the state creates the conditions for trust and coordination also among people with whom one is not related by kinship or otherwise acquainted.

Against this background, the argument of Hobbes and Locke is that the consequences of a stateless society (the state of nature) would be unbearable. In the Hobbesian picture, life in the stateless society is bound to be solitary, poor, nasty, brutish and short, because the conditions for mutual trust as a prerequisite for cooperation are not fulfilled (*Leviathan*, chapter 13). Misery would abound because the individuals would be unable to coordinate socio-economic relations beyond their families. The Lockean picture is much more positive, although the inconveniences that persons in the state of nature face cannot be solved within the state of nature without endless conflicts. In the end, the state is, according to this argument, the best solving-problem device that society can invent to guarantee peace and prosperity. The long-term benefits that the state brings for all underlie the reasons for developing a consent theory to justify it: eventually, people will come to agreements (contracts) about the necessity of a state. This is ultimately a consequentialist argument for political authority.

Once the existence and legitimacy of the state is assured, the next issue is about its legitimate extension. The liberal tradition only seeks to establish that a political authority is the best way to enforce rights and administer justice within a territory. But it stops there: the extension of a legitimate state is limited to *only* delivering justice and security. Any further extension is illegitimate. The reason is clear: the presumption of liberty always prevails and the state (and any extension of it) implies a limitation of liberty. Without counterbalancing reasons, any limitation is illegitimate. This is the rationale of what Robert Nozick (1974) calls the *minimal state*, which fairly represents the institutionalisation of the rule of law: formal and procedural principles that protect citizens not only from one another but also from the state by limiting its extension to well-specified duties.

Nozick's (1974) justification of the minimal state is, in short, this one: against the mentioned background of the modern society that is not based on relations of kinship, the need for security leads persons to take out security insurance. Market-driven mechanisms make it that the original competition between several protective associations winds up in one protective agency occupying a monopolistic position. This monopolistic agency best provides security and justice, resembling a state (indeed, an *ultraminimal* state). It becomes a *minimal* state by extending its services and its enforcement of security to *all* the people in a given territory. It thereby claims the monopoly of force (for criticisms, see Childs 1977 and Rothbard 1982, chapter 29). Nozick's justification of the state is practical in contrast to epistemic. There is no group that knows best how society must be shaped, what is the common good, or the right values to fulfil a genuinely self-determined life. The formation of the state comes about from purely practical circumstances, without it being intended as such or based on knowledge or morality (see Bader 2017).

In contrast to the consequentialist argument of Hobbes and Locke and the argument from practical circumstances by Nozick, Kant lays out a normative argument for the state that relies only on natural law without invoking any form of actual or implicit consent (*Metaphysics of Morals* 1795, part 1, §§ 41, 42, 45). The starting point of this argument is that every person is morally obliged by natural law, which grants her rights of freedom, to respect the rights of freedom of every other person unconditionally, that is, irrespective of whether she is willing to recognise these rights in a concrete case and agrees to respect these rights. Natural law legitimises the exercise of coercion against persons if they violate the rights of other persons.

The natural rights of freedom and the moral obligation for everyone to respect these rights take precedence over state power and the positive law that it establishes. But, according to Kant, because law is unconditionally valid, an authority is legitimate in which the power is concentrated to concretely enforce the law in a territory—that is, an authority that holds the monopoly in law making, law enforcement and judiciary in a territory.

Within the liberal tradition, there are thus three principled ways of justifying the implementation of the rule of law by a state that claims the monopoly of force in a territory: the normative one (Kant), the consequentialist one (Hobbes, Locke) and the one from practical circumstances

(Nozick). These justifications for the constitutional state are not based on the idea of a social contract, as contractual theories of the state have proposed. The moral basis of contractual theories is that individuals must voluntarily consent on being ruled by the state. But these theories have many issues and are insufficient to justify actual states. First, if this is a contract to which *all* parties consent, then there is no need for a state. The idea of a contract that involves everyone presupposes a degree of social coordination that would make the state superfluous. In that case, we would have a private law society: social interactions governed by contracts among the parties with arbitrators to resolve conflicts. There is then a social community under the rule of law, but no state that exercises a monopoly in law making, law enforcement and judiciary. Second, the consent-based theory supposes that everyone must agree on transferring rights to a sovereign, but no state contract has ever been concluded among the inhabitants of a territory in such terms. A contract is always based on actual—explicit or implicit—consent. There can be no obligation to consent. But there can be an obligation to submit to law. However, this obligation then exists independently of consent.

The idea of a state contract is nevertheless heuristically helpful. It reveals a dilemma that is inherent in the political implementation of the normative argument for the unconditional validity of *law* in the guise of a *state* that holds the monopoly of law making, law enforcement and judiciary: the contract establishes a state power that enforces what this power recognises as law. But the holder of the state power (the sovereign) is not itself a party to the contract. This fact is clearly expressed by Thomas Hobbes in the *Leviathan* (1651, chapter 17). *Hobbes' "Leviathan" therefore again marks a crossroads from which two paths start: the path to the modern constitutional state as well as the path to the modern totalitarian state. The latter path also branches off from the foundations of classical liberalism.*

According to Hobbes, the inhabitants of a territory conclude a—fictitious—contract among themselves by which they renounce their private interpretation and enforcement of natural law. They do so by appointing a sovereign who holds the monopoly on the use of force in the territory. The sovereign is charged with establishing and enforcing law in the territory, while the contracting parties are exclusively the inhabitants of the territory in question. The sovereign is thereby entitled to exercise the monopoly of force over the people in a given territory. Hence, not only the sovereign is given the whole power to interpret and enforce the law, but also the citizens have voluntarily renounced their own law interpretation and enforcement. The latter is a condition for the former.

The sovereign can be a single person (monarchy), an assembly of certain persons (aristocracy, oligarchy) or the assembly of all citizens (democracy). This does not matter for the legitimisation of the state. The crucial point is this one: when the monarch or assembly makes and enforces law by majority vote, it is not acting as a party to the fictitious state contract, but as an organ empowered by that fictitious contract to make law. This means that the *state power is not limited by the state contract*. Whatever it sets as law is considered law. The sovereign cannot be deposed or limited by the fictitious contract that establishes it. Not only Hobbes but also Kant make this point clear (*Metaphysics of morals*, § 49 note A). One can only withdraw from the state contract by leaving the territory in question or by starting a civil war. If citizens were to set their own judgement about what is the natural law, what rights they have and how these rights should be enforced above the judgement of the sovereign, then the monopolistic nature of state power is dissolved and we would fall back into the state of nature.

We can therefore discard the crutch of the fictitious state contract and formulate the decisive point in this way: there can be no secular legitimation to take action against the state power, because one would then set one's own judgement about what is right against the judgement of the state power. The power of the state, however, is there precisely to prevent everyone from imposing their own view of what is right, if necessary, by force.

The fact that the sovereign, who sets law and organises its enforcement, can also be an assembly and even the assembly of all inhabitants of a state territory (democracy), does not change this. In order to be able to take decisions, the assembly operates by majority vote, and be it a qualified majority, but not unanimously. But this means that coercion is exercised against the respective minority. Not all have agreed to decide by majority vote either: the state contract of all inhabitants of a territory among themselves, which establishes a sovereign, is, as mentioned, a fiction. When it comes to legitimising coercion, majority does not matter. A majority does not, by the mere fact of being a majority, have the legitimacy to exercise coercion against the respective minority. Natural freedom rights are not at stake in majority decisions: as the term "natural rights" expresses, they derive from human nature—the trio of freedom, reason and normativity as explained in Chap. 3, Sects. 3.3 and 3.4.

Consider the difference between contracting parties and the legitimation of the state. There are contracts between parties according to the pattern of protection against obedience. One party undertakes voluntarily to give protection to another party; in return, that other party commits themselves voluntarily to obey. But the duty to obey is limited in each case; the party to whom one undertakes to obey is a party to the contract in question. At least according to the concept, each party can terminate the contract at any time—that is, also the party who has committed themselves to obedience can terminate obedience and renounce protection without having to leave the place where they live. Whether this is always possible in practice is of course another issue. A cooperative is structured similarly: the members mutually undertake to follow instructions issued according to certain procedural rules (obedience) and receive protection from the community in return. However, each member can terminate this contract at any time and leave the cooperative without having to leave the place where they live.

The modern state is different. Its background is the mentioned problem of trust and coordination in social relations among people with whom one is not related by kinship or otherwise personally acquainted. Moreover, following the religious wars, the modern state is not held together by a shared religion or a collectivist narrative that could compensate the lack of trust and coordination among people who are not related by kinship. The basis of the modern state is the recognition of the freedom rights of every person. From this follows the obligation for every person to respect the freedom rights of every other person. However, the enforcement of this obligation by a state power with a monopoly on the use of force in a territory leads to a dilemma that is known as *Hobbes' dilemma*: on the one hand, the state power is only legitimate as a state under the rule of law; its mandate is limited to the task of enforcing the law in the sense of natural law, that is, to safeguard the freedom rights of every person and to protect every person from encroachment by other persons. On the other hand, the power of the state cannot be limited by the citizens in its judgement as to what is law in each case and what are the appropriate means in each case to protect the liberties of each person. It follows that the authority of the organs of the modern state in making positive law, enforcing it and settling conflicts (judiciary) is unlimited

Hobbes illustrates this dilemma with the *Leviathan*: the state power cannot be limited by the citizens in what it sets and enforces as law. Citizens could only set their own judgement about what is right against the judgement of the state power. But that means dissolving the state power. The *Leviathan* is not a monster, but the logical consequence of the premises (i) to recognise the freedom rights of every person and (ii) to enforce the corresponding obligation even without consensus by a state power with a monopoly on the use of force.

There are many illustrations of this dilemma. In order to effectively protect every person on its territory from violent attacks by other persons on life, limbs and property, the state authority would have to record everyone's whereabouts at all times, know all transactions etc. However, this would turn the constitutional state into a total surveillance state. Where is the boundary beyond which the state's implementation of the rule of law slides from a force that protects the liberties of every person against encroachment into a force that itself encroaches on the persons on its territory? Again, only the organs of the state power can judge this. The respective persons who feel threatened by these organs cannot enforce their own judgement on this boundary without thereby taking action to dissolve the state power.

The problem therefore boils down to the following: the state institutions are intended to protect the citizens. But once there is a state that holds the power of the monopoly of law making, law enforcement and judiciary in a territory, the functionaries of this power tend to extend their power under the pretext of improving the protection of every person in their territory against attacks by other persons. As we see, the state power nowadays grants protection even against viruses that people can spread, protection against climate change to which people supposedly contribute through their lifestyle and protection against holding and expressing views that (even if true) may hurt the feelings of some people. In order to provide this protection, the functionaries of state power must fully regulate the lives of the people in their territory. After all, everyone can contribute to the spread of viruses with any of their everyday activities, contribute in some way to the change in the world's climate or hold views that in some way may hurt the feelings of people who belong to groups that were somehow subjected to discrimination in the past. This is how the rule of law slides into totalitarian rule, as described in Chap. 2, Sect. 2.7.

Hobbes' dilemma concerns the internal political organisation of a society that is not based on kinship or o powers, as illustrated by ther relations of personal acquaintance. There is an analogous dilemma concerning the external relations of such a society. We propose to call this the *West Berlin dilemma*. This dilemma concerns the situation of an internally free society that is surrounded by hostile powers, as illustrated by West Berlin in the epoch between the end of the Second World War and the fall of the Berlin Wall. The threat posed by these hostile powers cannot be avoided by not interfering in their affairs: the mere existence of a free society in their proximity destabilises their system of rule—for example, by people under East Berlin rule trying to flee to West Berlin. In short, their system of rule is dependent not only on a monopoly internally but also on a monopoly externally, namely on the absence of alternative options.

The dilemma for a free society is therefore that, on the one hand, an extensive military and secret service apparatus is required as defence against the external threat. On the other hand, the structure of this apparatus, notably the command structure of the military, compulsory military service and secret agencies that are not bound to respect fundamental rights is diametrically opposed to the principles of a free society and the legal order on which a free society is based. One cannot expect people to leave behind their self-imposed immaturity, to use Kant's terms in his famous definition of enlightenment (see Kant 1983, p. 41) and to become responsible citizens that self-determine their lives in a constitutional, republican state and at the same time force them into compulsory military service and, in general, expect them to simply obey the orders of an extensive military apparatus.

The West Berlin dilemma concerns also the use of discretionary emergency law and the proclamation of a state of emergency in the case of an imminent external attack that threatens the existence of the internally free society. In general, every constitutional state also provides for the possibility of its own temporary suspension in the case of an emergency situation, in which the state organs are explicitly permitted to exercise coercion overruling even the natural rights of individuals. The classic case is the defence of the state territory against an external attack. In such a situation, coercive measures by state organs that override basic rights are permitted in the constitution in order to defend the existence of the very political order that secures these rights.

This presents another dilemma: on the one hand, it is recognised that in an emergency the state power must ultimately be equipped with unlimited command authority to effectively protect the people on its territory. On the other hand, it is up to the holders of state power to determine and thus to decide when an emergency exists. In the *Second Treatise of Government* (1690), chapter XIV, Locke grants the executive power the capacity of prerogative, that is, the right to discretionary exercise of coercion in the face of an emergency. The dilemma of the rule of law implemented by a state, hence, is that it must provide for its own abrogation in the event of an emergency and place the decision on the declaration of an emergency in the hands of those who have the power to suspend the constitutional rights in that event.

Consequently, we are back again to the same problem: once there is a state power that has the authority to impose a state of emergency with unlimited power, then the holders of this power tend to expand the scope of such emergencies. What is an emergency is decided by the political authority, and nobody else has the right to disobey the authority in its declaration of an emergency. The emergencies can then be of different nature. They do not have to be military wars with defence against an external attack on the territory of the state. They can also be metaphorical wars such as war against a virus or against climate change. Such emergencies are easily provoked. The state power can declare a wave of viruses a pandemic, relying only on infections that are produced by positive test results even if people do not manifest any symptoms of being sick and without there being excess mortality or systematic overcharge of hospitals.

The political philosophy of liberalism after Hobbes up to the present day attempts to limit the *Leviathan* de facto in order to prevent the abuse of its legislatively unrestricted authority. The vertical division of power through a federal state structure serves this purpose, along with what should be the right of each individual territorial entity to withdraw from the state. In Liechtenstein, for instance, even each individual municipality has the right to leave the state of Liechtenstein. This is further served by the horizontal separation of powers between a legislative, an executive and a judicial authority at all levels, the conception of which goes back to John Locke's *Second Treatise of Government* (1690) and Charles-Louis de Montesquieu's *De l'esprit des lois* (1748). This is also served by a sophisticated system of "checks and balances", as implemented, for example, in the constitution of the United States.

Above all, however, the idea of the republican constitutional state, which Kant elaborates on in particular in the *Metaphysics of Morals* (1795, public law) and the treatise *On Perpetual Peace* (1795, first definitive article), is the answer to Hobbes' dilemma. The republican constitutional state brings the idea of limiting power through the separation of powers together with the principle of participation. The republican idea is the idea of collective freedom as freedom from foreign domination. Rule is self-imposed rule in which the ruled participate.

From the constitution of the United States in 1787 onwards, the concept of the republican constitutional state was implemented in the form of liberal democracies: the participation of the citizens in their governance takes place by law making through an assembly. All inhabitants of an area can be entitled to vote in this assembly, or the right to vote is restricted by conditions that define citizenship rights, such as owning land or paying taxes, or being prepared to give one's life in defence of the state. The latter was originally the case in the Swiss *Landsgemeinde*, where one acquired the right to participate in the decision-making on public affairs by being willing to serve in the military. Therefore, in some places, citizens appeared in the *Landsgemeinde* with their arms to show that they were entitled to vote because they were prepared to defend the country by putting their lives at risk if necessary. If an assembly of all citizens is not possible because of the size of the territory (e.g. as in the United States), the participation of all citizens in political decisions takes place through the election of representatives.

Participation includes not only the right to vote but also the right to stand for public office. The allocation of these offices is subject to the principle of rotation: each office is limited in its power and is held by the same person only for a certain period of time. Holding political mandates is not a profession, but a service to the community that is exercised for a certain period of time in addition to the profession that one has.

However, all this does not change the fact that this is only a pragmatic solution. The Hobbes dilemma and the West Berlin dilemma cannot be resolved within a state that claims the monopoly of force. Once a state power with such a monopoly exists, there is a tendency for its power to expand, however this power may be structured and whatever mechanisms may exist to control it. Even though the pragmatic solution to these dilemmas has proved viable over long periods of time, all these mechanisms, including the separation of powers and the federal state structure, can fail. We have seen how easily all these mechanisms can break down in the staging of the coronavirus pandemic.

Even the courts up to the constitutional courts failed in the application of judicial diligence: in general, they did not examine whether the suspension of fundamental rights in positive law and the regulations of the authorities on the so-called corona protection measures were lawful and proportionate in the sense of the first articles of the state constitutions, which guarantee fundamental rights stemming from natural law. The court rulings—for instance, the ones of the German Federal Constitutional

Court (Bundesverfassungsgericht 2022)—simply referred to positive law and to what was presented by the state authorities as alleged science in order to legitimise this positive law. They thus deprived positive law of its foundation in natural law.

Far-reaching restrictions on human rights can also be approved in referendums, as we saw in the votes on the COVID-19 law in Switzerland in June 2021, November 2021 and June 2023. The state power can influence such votes, for example by providing financial incentives (COVID-19 aid financed by state debt, June 2021 vote) or by spreading misinformation (vaccination stops the spread of the virus, November 2021 vote). The frightening thing about all this is the following one: *if something relatively harmless to the general population as the coronavirus waves can have the effect that the mechanisms for limiting power within constitutional states fail, then anything can have this consequence.* However, these mechanisms did not simply fail on their own. Other factors contributed to this, which we shall bring together in the next two sections.

To sum up this section, the republican idea of implementing the rule of law in a constitutional state whose organs hold the monopoly of law making (legislative), law enforcement (executive) and judiciary in a territory leads to the following dilemma that already Thomas Hobbes illustrates: the one horn of the dilemma is Hobbes' argument that in order to avoid a state of nature that is characterised by the uncontrolled use of violence because everybody enforces their own interpretation of natural law and natural rights, a sovereign has to be erected that holds the monopoly of law making, law enforcement and judiciary; if there were not a sovereign with that power, uncontrolled violence would reign. The other horn of the dilemma is that the state power cannot be limited by the citizens in what it sets and enforces as law. Citizens could only set their own judgement about what is right according to natural law against the judgement of the state power. But that means dissolving the state power and returning to the state of nature.

This dilemma about the internal political organisation of a free society under the rule of law is complemented by a further dilemma that may concern its external situation and that can be dubbed the West Berlin dilemma: an internally free society that is surrounded by hostile powers requires a military apparatus for its defence whose organisation is opposed to the principles of a free society. In particular, the state authorities are granted the use of discretionary emergency law and the proclamation of a state of emergency as defence against an imminent external threat in which the basic rights of the citizens are suspended. However, once there is a state power that has the authority to impose a state of emergency with unlimited power, then the holders of this power tend to expand the scope of such emergencies.

These dilemmas are inherent to the whole liberal tradition. They are not about deficiencies of human beings that could be resolved by re-education, envisaging the creation of new human beings as in the totalitarian ideologies from the nineteenth century onwards to this day. The liberal tradition accepts humans as they are. But this implies that in a stateless society, people are prone to resort to the uncontrolled use of violence because there is no authority that is entitled to settle the interpretation of natural law. However, once there is such an authority, there is the risk that the persons who hold the corresponding power abuse it in the guise of arbitrary employments of their power of law making, law enforcing and judiciary.

4.4 From the Welfare State to Welfare Totalitarianism

The rule of law (as implemented in the constitutional state) aims to safeguard the negative rights that derive from natural law. However, in virtue of the monopoly of law making, law enforcement and judiciary, the organs of the modern constitutional state have unlimited power in judging what is appropriate to protect these negative rights. This monopoly tends to abuse, namely to an unlimited extension as to what this protection should consist in. From the nineteenth century on, constitutional states offer not only protection against unwanted external interference by other persons but also protection against the central risks in life such as risks from old age, poor health or unemployment. The constitutional state thereby turns into the welfare state. The welfare state is characterised by not only safeguarding negative rights but also fulfilling positive rights to certain entitlements (for instance, social benefits).

To see more clearly the logic of this shift, consider the following. Under the rule of law, coercion against a person is legitimate if and only if the person in question violates the negative rights of other persons or sets out to do so. Enforcement of contracts among individuals is also covered by this task insofar as the non-fulfilment of a contract by one party violates the other party's property rights. Positive rights are acquired and fulfilled through voluntary agreements between members of the civil society. In many cases, thus acquired positive rights seek to satisfy a need. For instance, I offer the use of my labour force to others in order to earn money that enables me to buy food and services. Many social and economic relations are a passing around of positive rights and obligations—if I make a contract with others in putting the use of my labour force at their disposal, my labour force gives me a right to dispose of some amount of money, while the money I spend in the supermarket gives me rights to enjoy some goods. Under the rule of law, it is not the task of states to interfere with how persons pass around these positive rights as this is carried out through voluntary agreements following individual goals and purposes.

When the state's power shifts from enforcing negative rights to fulfilling positive rights, the road to welfare totalitarianism begins—that is, the road to serfdom as Hayek (1944) puts it. First and foremost, as soon as the state fulfils positive rights to certain entitlements, its organs have to use coercion to acquire the means to fulfil positive rights and satisfy the corresponding entitlements. That is, they have to violate the basic principle of natural law according to which coercion is justified if and only if a person acts in such a way that she encroaches upon other persons' negative rights. The transition from the constitutional state implementing the rule of law to the welfare state thus is characterised by the state organs resorting to coercion to fulfil entitlements rights that they grant to the people on their territory (and often also beyond, such as granting entitlement rights to foreigners in the case of immigration).

Positive rights are typically tied to the satisfaction of needs. However, needs have no limit. By way of consequence, the state power can be extended without limit to fulfil in principle countless positive rights. For instance, people have the need for electricity. This need can be satisfied through private-to-private agreements. But if such a need is transformed into a state-enforceable positive right, then the rulers are now allowed to coerce others in order to fulfil such a right. They can, for instance, force other persons to provide free electricity, or can subsidise with taxpayers' money the access to this service.

The republican, constitutional state thereby shifts from setting the framework in which the social and economic relations among the people develop through voluntary agreements to pursuing a common good in terms of what its organs define as the people's welfare. Under voluntary interactions and contracts, individuals exercise their positive freedom as self-determination in shaping their lives as they see fit under the circumstances they face. Under the rule of law framework, it is up to the individuals to find the way to satisfy their needs through voluntary cooperation. In this sense, positive freedom is abstract and compatible with indefinitely many ways to shape one's life. However, when the welfare state steps in and seeks to guarantee positive freedom, to coercively fulfil individuals' needs, many undesirable and worrisome consequences follow.

To start with, the state organs have to set what they consider as providing the material conditions for self-determination—for instance, securing a universal basic income through entitlements to social benefits. Yet, this is a coercive mechanism because the fulfilment of unconditional positive rights requires resources that states arbitrarily take from individuals and allocate differently. This goes against both positive freedom and negative rights: voluntary interactions, which represent individuals' preferences, are altered and property rights are violated. Furthermore, welfare states fix a common good that represents the rulers' view of self-determination and that is to be collectively realised through coercion. This can take many forms. Public paternalism, for instance, is a way in which rulers can override people's individual freedom and aim to realise a superior good (see Dworkin 1972, 2013). Forced collectivism is another, in which rulers see individuals as cogs in a broader system that must aim for a common good. Thirdly, the protection in the face of an alleged emergency is a way to justify far-reaching interventions, pretending that individuals are unable to deal with the emergency through their voluntary interactions and have to acquiesce in the central planning of their lives.

One of the main arguments in favour of welfare state interventions is that they are necessary to avoid or correct alleged market failures. The argument is that in voluntary agreements, individuals behave selfishly, without caring about other individuals' welfare. This generates a massive number of negative externalities and socio-economics issues that need to be centrally intervened upon and regulated.

Negative externalities are negative effects on third parties that result from contracts among people. A classic example is this one: people who live on the upper part of a river agree to build a factory whose operation pollutes the river's water. The inhabitants further down by the river do not benefit from the factory and are not party to the contract to build the factory. However, they are exposed to the pollution of the river. For example, they can no longer safely swim in the river or organise tourist activities on the river. The pollution of the river is a negative externality for the residents on the lower part of the river. But there are also positive externalities, as Jörg Guido Hülsmann emphasises in his most recent book (2024, chapter 6): for instance, if the owner of a pub hires a security service to control the entrance of the pub, the people living in the neighbourhood of the pub profit also from this protection (e.g. criminals are deterred) without being party to the hiring contract of the security agents.

Furthermore, it is alleged that if social interactions are based on voluntary agreements only, the resulting free-market societies allegedly produce deep inequalities and poverty, which in turn generate numerous negative externalities (e.g. slums and high crime rates). The solution then is to regulate the economy in such a way that wealth is redistributed more fairly, counterbalancing market-driven inequalities and negative externalities. In a more general fashion, the idea is that markets have a natural tendency to undersupply (or oversupply) goods that entail positive and negative externalities; therefore, widespread intervention on markets on behalf of welfare states is necessary to restore equilibrium and fairness.

There are three big mistakes in this reasoning. The first one is the unwarranted assumption that individuals are not social beings. Individuals are not plainly selfish, but they are in average selfish *and* altruist; furthermore, both these characteristics often go together. As social beings, humans need to interact with each other and to form communities: thinking and acting is bound to social interactions. Thus, any voluntary economic exchange is such that both parties are better off as a result of the exchange. Otherwise, they would not willingly agree to the exchange. Suppose that I own five apples and my neighbour owns five breads. To exchange an apple for a bread benefits us both. The exchange thus is both selfish and altruist. The reason is that owning more apples has only little use for me, whereas owning some other stuff—such as bread—that I do not possess as yet has significant use for me. The same applies to my neighbour for whom owning more bread has little use, but owning an apple big use. Thus, both parties are better off as a result of the exchange. Generally speaking, given the different capacities of individuals and the different circumstances they are in, the division of labour through voluntary exchange in a free market is profitable to all parties of such an exchange. They are all better off because they are free to agree to all sorts of exchange. As Jeffrey Tucker (2019) puts it, "The market loves you". Free economic exchange promotes the well-being of us all. It enables us to achieve our goals.

Second, since humans are social beings, they furthermore join voluntarily and spontaneously to form social communities in which they support each other and thus promote conditions for freedom to unfold as self-determination: families, circles of friends, interest groups, diverse cultural and religious communities, cooperatives that organise the division of labour and mutual support among their members etc. One of the negative consequences of welfare states is that they weaken many of the voluntary and spontaneous social bonds that individuals establish to pursue common aims and thereby also providing free goods (see Hülsmann 2024, part 3). In particular, they crowd out voluntary associations such as cooperatives that provide protection against risks in life such as old age, poor health, and unemployment.

The third mistake is the unjustified assumption that alleged market failures call for state intervention such as state regulation or even state ownership. To begin, the idea of "market failures" is unclear as it supposes an objective standard to which market results can be compared. How do we know that a specific market result is a failure? This assumes that someone knows which is the "correct" result and then is entitled to rectify the failure. However, this is a misconception. Markets are processes of learning and discovering. There is no external viewpoint from which to judge the results. This does not mean that markets do not fail; it means that we do not have a viewpoint from outside the market to determine if they fail. As repeatedly mentioned in this book, the state authorities lack the knowledge to correct alleged market failures through interventions. In addition, while market failures are emphasised, government failures are ignored. As notably Mises (1949) points out, any state intervention to counterbalance a negative externality leads to new negative externalities that call for further state interventions and so on. This is Hayek's (1944) road to serfdom: the correction of government failure consists in more government intervention, ending up with pervasive government control.

To avoid this spiral of interventions, concrete cases of negative externalities call for alternative mechanisms that can correct the situation depending on the circumstances: for instance, settlement by a judge or a mediator that is tailor made for the case at hand instead of central state regulation (that can in any case not do justice to all possible circumstances); a system of private compensation or insurance; a clearer determination of property rights in each case that establishes where, when and to which extent compensation is due when demanded etc.

Thus, to come back to the above-mentioned example of a river being polluted by the operation of a factory, the judgement of a neutral authority—an arbitrator—may be required to bring about a balance. The equalisation can, for example, consist of imposing conditions on the operation of the factory or awarding compensation to the residents down by the river. By contrast, it would not be permissible within a legal system that implements the rule of law to argue that the operation of factories is a common good and therefore those affected must accept negative consequences out of "solidarity"—or, conversely, to argue that everyone has a right to live in an untouched environment. Every technological innovation has an impact on the natural environment, which is in itself in constant evolution even without human interventions. If human interventions result in negative externalities that can be proven by objective facts, a judge or a mediator is called for to bring about a balance in the case at hand.

Another argument in favour of welfare state intervention is to avoid free riders who benefit from a commonly sustained enterprise without contributing to it. A free rider, it is argued, benefits from positive externalities that the existence of public goods generates. Since it is unjust that free riders benefit for free, they are forced to contribute to the welfare state. This argument is also flawed. To begin, if the problem is the existence of free riders, it is unclear how another free rider with the monopoly of the force on a territory (as the state is) can be of any help. State officials are by definition free riders since their only income comes from taxation, that is, from the fruit of other people's labour: they live at the expense of others. It may be argued that they serve others by providing state-run services. But this paradoxically creates the conditions for the most successful way to live at the expense of others, since the welfare state can be used for one's own interests.

The state alone has the power of legitimate coercion. This generates negative incentives for people at power of extensive states. It furthermore incites influential individuals to use this power for their own interests. This means that as soon as state officials deviate from enforcing equal rights for all and set out to intervene in individuals' lives and voluntary interchanges, they open the door to lobbying. Representatives of various special interests can then best advance their interests by pretending to stand up for the common good and persuading state officials to take measures in the name of this common good that, in substance, only serve their respective special interests. It is an illusion to believe that there is an unselfish elite that will govern justly only by considering the public good. Real governments involve actual people, with their virtues and vices, giving incentives to unleash their vices rather than their virtues.

The overgrowing of the state to become a welfare state (welfare statism) necessarily requires some degree of intervention upon the economy. Otherwise, the state would lack the resources to sustain its overgrown structure. However, this intervention erodes economic relations, leading to the corruption of the economy. An economy based on free trade, free contract and the division of labour is the way to progress in productivity and the optimal satisfaction of everyone's needs: everyone contributes to what they can do best and in return receives services from others to satisfy their own needs. The division of labour can be organised on a voluntary cooperative basis or under a capitalist wage system. Without state intervention on the economy, there will probably be an interplay of various forms of organising economical activities. From a purely economic point of view, the capitalist form of economy may be more efficient than the cooperative form. However, the latter is attractive because of the protection and the community it offers its members.

Welfare states are then in tension with free-market societies. In free markets, some form of transmitting information, a structure of incentives and mechanisms of learning is required. As mentioned before, this does not entail that market failures do not exist, but that they can only be seen from within and be fixed by market-driven mechanisms. After all, any well-behaved market entails learning from mistakes. By contrast, to try to compensate for market failures through the intervention of state power is to cast out the devil with the Beelzebub. If there are state functionaries who intervene in the economy by using the state's coercive power, then not only there are distortions in markets but also it becomes attractive for economic actors to influence politicians in the interests of their companies' profits, generating further distortions.

When the state tries to steer markets, capitalism becomes *state* capitalism, which is bound to degenerate into *crony* capitalism. Welfare states and crony capitalism go hand-in-hand. Instead of customers, entrepreneurs and their associations competing to provide better goods and services to the best price, crony capitalism creates the incentive structure for lobbying and colluding with the state to gain a hegemonic position in a market. This usually translates into profits still being private, but risks and losses put on taxpayers' burden. The paradigmatical case of this is the massive amount of taxpayers' money that is directly transferred to big banks and companies in the form of state bailouts. A state-created incentive structure is thus generated, in which big banks and companies run into high risks because they know that in case of failure, they can ultimately rely on bailouts. Artificial private monopolies are also paradigmatic cases: regulations, or direct state favours, greatly benefit some companies at the expenses of consumers and other companies. Protectionist policies usually have this effect, as do patent laws.

A very recent and particularly appalling case of welfare statism—or even welfare totalitarianism—and crony capitalism was the vaccination campaign within the corona regime. Pharmaceutical companies developed and sold vaccines to governments under confidential contracts that have not been disclosed to this day. The liability for the product was waived by most governments, eliminating any requirement of compensation in case of severe counter-effects or even deaths. This basically means that when the product is out in the market, it can be a complete failure, but the pharmaceutical companies remain at save since governments waived liability. To add insult to injury, governments massively bought vaccines with taxpayers' money and forced individuals to get vaccinated to recover what used to be basic civil rights as compensations for obedience (e.g. to enjoy dinner at a restaurant). The consequence is clear: pharmaceutical companies could make gigantic profits thanks to the use of coercion by the welfare states. The welfare states pretended to protect citizens from a risk, artificially created a demand, took taxpayers' money to satisfy it and forced them to consume the products. It is worth recalling Kant's words as quoted at the beginning of Chap. 4, Sect. 4.2:

> But if everyone were free without law, nothing more terrible could be conceived. For each would do with the other as he wished, and so no one would be free. (Kant 1979, p. 1320, our translation)

This is exactly what happens under crony capitalism: some are free without law and to do with the others as they wish: economical influential individuals make use of the monopoly of force of the state for their profit without being liable for their actions. While in free-market capitalism individuals compete with one another in a self-regulating mechanism, crony capitalism creates state-fuelled monopolies and pervasive abuse on their behalf.

It is an illusion to believe that welfare states protect the worse off. By contrast, voluntary associations within civil society can enforce the interests of various individuals, such as worker unions, consumer associations, trade unions, non-profitable organisations, charities etc. They are not granted by states, but they emerge spontaneously from the civil society. If some people with common interests consider that they are at the mercy of a more powerful group with opposite interests, then the appropriate reaction is to join forces. As Socrates already responds to Callicles' speech in Plato's *Gorgias* (488c–489a): the many, when they join, are more powerful than the few strong. They can, for example, refuse to cooperate with the powerful, stop buying their products and services, or produce their own products and services cooperatively. No central planning, no state regulations, no state protection is needed: voluntary associations historically precede their statist forms.

Welfare states do not only regulate the economy but also tend to wrap this regulation into a narrative that holds society together. In the nineteenth century, the prevailing form in which this occurred was the formation of nation states. Nation states, however, tend to be repressive internally and aggressive externally. Repression internally consists of forcing everyone under the banner of the nation in the name of an alleged national interest or national common good. This amounts to, for instance, corresponding regulations of the economy to foster national industries, protectionism for some economic sectors to avoid foreign competition and a social security system based on coercion. External aggression consists at least in the tendency to assert and expand the power thus constituted in competition with other states and to build up armies accordingly, to promote arms production etc. The result is what is known as the welfare-warfare state. The imperialism that the nation states promoted in the nineteenth century ended in the First World War. The First World War in turn paved the way for the totalitarian political systems of the twentieth century. It is no coincidence that the two states that come closest to the ideal of a republican constitutional state are not nation states: Switzerland and the United States. However, because of its size, the United States has also allowed itself to be drawn into a foreign policy of aggression against which, however, there has always been opposition on the part of libertarian currents at home.

State interventions lead to corruption not only in the economy but also in other fields of society. An open, free society depends on a public space in which the exchange between the different ways of life, religions, ideas etc. of the people living together in a territory takes place in a mutually enriching way. Modern constitutional states have all developed, to a greater or lesser extent, in such a way that the state power not only provides the framework for this public space by ensuring the right of everyone to live freely but also organises and shapes this space, controlling how these non-commercial relations and exchanges take place. This is done primarily through the state's monopoly over the most influential media in the twentieth century, namely, radio and television. Likewise, the state operates education and science largely in its hands. Even to the extent that there is private ownership in these areas, it depends on central state regulations and approbation and thus on compliance with the state organs.

Even though the main argument in favour of this control is that it is the best way to avoid that monopolistic private groups can gain a dominant position in these areas, reality is that if the most influential media, the educational and scientific system are controlled and regulated by the state, this impedes competition in these areas and thus leads to impoverishing their quality. In parallel, special interest groups can now collude with the state to not only obtain economic benefits but also impose their worldviews, moral concepts, political agendas etc. In short, the supposition that pluralism is best guaranteed by the state power is just an illusion and has led to the opposite resulted. In our days, the forces driving actually existing postmodernism use the public media, the state-organised education system and the state-funded science as their central instruments. This instrumentalisation is made possible by the concentration of power in the hands of the state organs, which is linked to the monopoly over influential media, the educational system and science. This monopoly does not prevent special interests from asserting themselves, but on the contrary promotes this development.

In particular, a scheme of state-funded education and science tends to foster political scientism. There always are scientists and intellectuals who are prone to political scientism. If there is a broad state monopoly over influential media, education and science, then the goal-oriented strategy for political scientists is to seek to hijack these institutions. If political scientists agree on a narrative or sequence of narratives, as in the corona, the climate and the wokeness regimes, then they can succeed in taking over these institutions through their coordination and determination. The dominance of such an ideologically homogeneous group in influential media, education and science is even more dangerous than the worst form of economic corruption. Those who dominate influential media, education and science can influence the use of language and reinterpret concepts, depriving people of their self-determination and faculty of judgement without realising what is going on—or realising this only when they find themselves already in an ideologically closed society.

In an open society based on voluntary interactions, the demands that individuals can make are limited by the services that others are willing to provide out of their own free will. All parts are aware of the benefits and costs of their decisions and interactions. Not so with welfare states: who has to pay for the fulfilment of claims remains nebulous, as the resources come from taxpayers' money and look as if they were obtained out of nothing. However, as there is no free lunch, the discretionary transference of resources by welfare states necessarily makes some individuals worse off and others better off. There is yet a well-marked limit for this, since there are obvious limits beyond which high taxation becomes unpopular. Welfare states can also finance themselves through debts, but their payment also relies on taxpayers' money. In any case, the limit is flexible, but clear.

It would therefore be ideal for welfare states to find a way to afford states' expenses without directly relying on taxpayers' money, or international debt. In some sense, everything boils down to getting money, somehow. What if it is possible to create money out of nothing? If welfare states have not only the monopoly of creating money but also the power of creating it at will, then they would not need to convince voters to pay higher taxes to afford their expenses. They could just create the money to pay off their expensive costs and sweep under the rug the dirt and undesirable consequences. The story of how welfare states devise the mechanisms to create money out of nothing (or fiat money) leads directly to the economic form of actually existing postmodernism.

State officials have always tried in history to usurp the monopoly over money on their territory. Government control of money can be traced back to the Ancient Age, when it was largely based on the coinage prerogative. It is not until the eighteenth and nineteenth centuries that a robust banking system consolidated itself, assuming further functions in issuing, lending, depositing and transferring money. The banking system was originally a private institution that provided the resources to finance the expansion of industry and infrastructure throughout the Western world. But it is the emergence of modern central banks out of a largely private banking system that was the main target of state officials. Among the functions of modern central banks are the regulation of national currency, the monetary assistance to government, the monopoly in the distribution and issuance of currency and the task of being a lender of last resort. By the end of the nineteenth century, most central banks in Europe were partially or completely organs of the state. In the twentieth century, central banks were also created in America, most of them as organ of the state, too.

The real economy, however, is not about money. It is about people and the goods and services they produce. Adam Smith writes in *The Wealth of Nations* that "the sole use of money is to circulate consumable goods" (Smith 1976 [1776], book 2, chapter 3, p. 339). Goods represent wealth. Humans are extremely talented at creating wealth. Wealth creation is the consequence of individuals working together in the division of labour, where they can deploy their unique skills and aptitudes in a collaborative enterprise. In this framework, money is an obvious consequence in the production of wealth, being an efficient means to interchange goods and services. It just works as an agreed upon measure among individuals to facilitate the exchange of production. Therefore, it is a truism that individuals do not want money on its own, but the goods or services money can be interchanged for.

Money being a measure, the expansion of the measure cannot produce further goods and services. For this reason, it was for a time known that one of the central virtues of money as a measure should be its stability. Investors and consumers alike need to presume a stable measuring of value to perform any interchange. This explains why money's evolution always sought for a stable parameter to define the measure. Gold and silver emerged as the definer of the measure because, as John Stuart Mill put it, "Of all commodities, they are among the least influenced by any of the causes which produce fluctuations of value" (Mill 1885 [1848], book 3, chapter 2, § 2, p. 291). In the nineteenth century, there was a gold standard linking the various national currencies because stability was crucial for producers and traders of actual wealth. A fixed standard to which currencies are pegged also limits the capacity of state officials to issue money.

For this reason, Western states suspended the convertibility of their currencies to gold many times to finance wars and war reparations. In particular, most belligerent states suspended the gold-exchange standard in 1914 when the First World War broke out. Post-War Germany could not even come back to it as it had to pay war reparations and war loans that it could not afford. It then issued a huge amount of non-convertible marks (fiat money) to face its expenses, thereby provoking one of the most terrible hyperinflations of the twentieth century. That notwithstanding, with its ups and downs, it was possible to maintain a partial gold standard across states after the First World War. For the period after the Second World

War, this partial gold standard was anchored in the Bretton Woods Agreement of 1944: the US dollar remained defined as 1/35 of a troy ounce of gold. All other currencies were tied to the US dollar by fixed exchange rates. Even though individuals were left out of this market, government-run central banks of other countries could exchange their dollars at the US central bank for gold at the defined ratio of 35 dollars for one troy ounce of gold (and likewise purchase dollars for gold at this rate).

It is important to think about this point carefully. Money is just a medium to trade goods and services for other goods and services. It is extremely beneficious and efficient to count on a stable measure that facilitates transactions. When money issuing becomes the monopoly of the state, it is essential to rely on some external element to which the exchange rate is pegged to limit arbitrariness in the exercise of power in the guise of the creation of money. Of course, governments can always devaluate against the fixed standard, as the United States did repeatedly, but at least a gold-backed currency can serve as indicator of the stability of prices and overall wealth. There is nothing special about gold beyond its stability. It is not about gold per se, but about having a limit that keeps governments' arbitrariness at bay.

What, then, is the relation between money, postmodernism and welfare states? As we pointed out repeatedly, reason is a limiting capacity: the use of reason can limit the exercise of power in society by means of the rule of law, it can limit the scope of dogmatism in science through disciplined scepticism etc. Postmodernism looms when the limiting power of reason breaks. In the 1960s, the dollar's peg to gold came apart at the seams because in domestic policy the United States' government wanted to satisfy more and more welfare demands without creating wealth (Johnson's "great society"), and in foreign policy, it used military means to assert claims to power (Vietnam War). In this way, the money supply needed to maintain the welfare-warfare state expanded far beyond the gold coverage (by 1970, it went down from 55% to 22%). As a result, the gold holdings of the Federal Reserve threatened to melt away: more and more dollars were put into circulation without the underlying real value increasing. The consequence was that this real value dwindled as foreign central banks increasingly sought to exchange surplus dollars for gold.

Faced with the choice of aligning the US government's claims to power with the reality to which the currency was pegged (i.e. either acquiescing in a big devaluation or a severe reduction of government expenses) or creating the illusion of reality to sustain these claims to power, President

Richard Nixon chose the latter. On 15 August 1971, Nixon "closed the gold window" by suspending the convertibility of the dollar to gold. In practice, this was the end of the pegged dollar-gold system. By 1976, the Bretton Woods System finally collapsed giving birth to a floating exchange rate system with fiat currencies. In 2002, the first president of the European Central Bank, Willem Duisenberg, even praised the Euro as the first currency in the world that is not backed by anything since its introduction (Duisenberg 2002; see Bagus 2010 for an argument why this constitutes "The tragedy of the Euro").

15 August 1971 ushers in the first, economic phase of actually existing postmodernism. Welfare states now have the unrestricted licence to issue money to afford any of their expenses, without any correlated increase in wealth (i.e. goods and services produced by people in an economy). This amounts to the creation of an illusory reality: welfare states have the capacity to generate the illusion that more goods and services can be acquired by issuing money out of nothing. But this is just a *postfactual* reality at work in the monetary system: there are no facts that determine and, therefore, limit the exercise of power by welfare states to issue money.

It is worth coming back to the original reasoning to realise the devastating implications of this. Fiat money undermines stabilisation as money can be created at will by state officials to be exchanged for real goods and services that individuals produce in an economy. The underlying logic of having a monetary system now breaks down: instead of using money as a stable medium to interchange goods and services for other goods and services, fiat money serves now as interchange for other goods and services without anything real backing it. This is a perfect expression of what postmodern thinkers call simulacra: a replacement of reality with its representation. Fiat money is just postmodern simulation in economics: copies of wealth that no longer represent wealth, but that are just empty copies.

Welfare states have all the incentives to increase and feed the fiat money system. The creation of new demands can now be satisfied by issuing non-backed money, gaining votes in elections. The socialisation of risks and losses can now be easily implemented since welfare states can bail out companies and banks without directly taxing the population. The increase of social benefits to certain groups of the population can be more easily carried out since states' expenses can grow without control. Of course, all this has terrible consequences because reality eventually speaks out. Nonetheless, the mechanism is short-term effective: welfare state can create the illusion that everyone is wealthier by putting the printing press at work.

In fact, however, the creation of fiat money and the state debt to which it leads, requiring the artificial suppression of interest rates by the states' central banks, again confirm that welfare states do not promote social equality. To consider just one figure, from 1860 until the 1980s, an US American always had to work an average of between 15 and 40 hours to be able to buy a share of the US stock index S&P 500 (mean 30 hours). Since the end of the 1980s, this value has skyrocketed: at the beginning of 2020, 120 average hourly wages were required to buy the S&P 500 once (see Baltrusaitis 2020). The unlimited multiplication of fiat money is a major factor in driving up the prices of capital goods such as shares and real estate, without wages keeping pace with this development: this development is not based on the growth of productivity, but only on the growth of the money supply. In other words, the fiat money system is anti-social. It benefits the owners of capital at the expense of workers. It thereby raises global income and wealth inequality through monetary interventionism of the states.[2] State control of the market through the fiat money system prevents the win-win situations that arise in the free market through voluntary cooperation. Accordingly, replacing state-controlled fiat money with private money as a market good is the means to end the devaluation of labour against capital.

The first phase of actually existing postmodernism is not totalitarian. This, however, changes with the corona, the climate and the wokeness regimes. With these regimes, actually existing postmodernism enters its second, totalitarian phase: *welfare totalitarianism*. Under these regimes, unlimited and thus totalitarian control of people's lives takes place under the pretext of unlimited and thus totalitarian protection from all sorts of risks, including the risks of virus transmission, risks in the change of climate and risks to one's subjective well-being from the expression of views that one may not like. This second phase presupposes the first one: without the extension of state power over society and the economy—the steering of markets by changing the allocation of resources, the monopoly on science, education, health as well as the monetary system—welfare totalitarianism could not sustain itself. Without the prerogative of issuing non-convertible currencies, welfare states could not afford their expenses.

[2] See in particular Hülsmann (2008, 2014), Bagchi et al. (2019), Othman et al. (2020), Polleit (2022a, chapters 6, 8, 9, 16–20).

The immense costs of the corona, the climate and the wokeness regime would then be obvious to everyone: everyone would feel the consequences in their wallets through the increase in taxes and duties that would be required to finance such an overextended, totalitarian welfare state.

The corona, the climate and the wokeness regime constitute a transition from welfare states to welfare totalitarianism for at least the following reasons: first, they openly trespass the limits of the rule of law. Second, they see the state as the guarantor of universal positive rights. Third, they claim to possess scientific expert knowledge of the common good. Fourth, they claim to possess expert knowledge of emergencies and of an objective assessment of risks. Finally, the result is a pervasive violation of basic, negative rights, along with systematic failure in fostering social cooperation, peaceful social relations and prosperity for societies as a whole.

Welfare states become totalitarian when the use of coercion (or the threat thereof) aims at controlling all aspects of life, including private life. Karl Popper (1945) refers to totalitarianism as a commitment to holistic, utopian social engineering as it aims to transform society and human beings according to a collective plan, overriding any individual plan. In this sense, totalitarian regimes always seek to realise some common good, some ideal of self-realisation and perfection. To realise the totalitarian rulers' view of the common good, all the instruments of state power are available—after all, the perfection of the common good justifies any means. But the means should not be confused with the defining traits. It is true that the twentieth century witnessed infamous examples of bloody, ruthless totalitarian socialist regimes both of the national socialist type and of the communist type. However, a state does not need to be as bloody and ruthless as these regimes to be qualified as totalitarian. What matters is trespassing the boundaries of the rule of law to realise a particular view of the common good through unlimited regulation of people's lives, including what used to be their private lives, turning rights into privileges, reversing the burden of proof etc., as described in Chap. 2, Sect. 2.7.

The welfare component in welfare totalitarianism relates to the fact that this is done under the pretext of protecting people from all sorts of risks that can arise in life. The postmodern component is that the big narratives that postulate an ultimate goal of history are replaced with a sequence of small narratives that each postulate a partial common good such as health protection, protection from climate change, or "social justice". That notwithstanding, welfare totalitarianism preserves many of the characteristic traits of totalitarian regimes, such as the control of education, control of the economy, control of the population through propaganda and the use of alleged scientific knowledge to justify itself and its policies. This last point is crucial because it is what connects welfare totalitarianism with political scientism—an elite of rulers seeking to use coercion to shape society conforming to its view, using modern science to justify the means and the end.

At the roots of welfare totalitarianism is what we call the *common good argument*. It assumes that there is a knowable common good. People ought to contribute to the common good. Informed actions perform better than uninformed or misinformed actions when it comes to contributing to the common good. Uninformed and misinformed actions should therefore be deterred and informed actions be promoted. In order for people to be informed, they should then defer to experts who know the common good and how to promote it. Guided by expert knowledge, governments therefore have the right to intervene upon people's ways of fathoming their lives, especially when they incur in uninformed or misinformed actions that do not contribute to the common good.

This argument is very powerful in our days. But it undermines the basis of modern science and the rule of law, leading to the collusion of political scientism and incipient forms of welfare totalitarianism. To begin, it claims that there is *knowledge* of the common good. The rulers are justified in extending the state's coercion *because* they aim for the common good which they know. Scientific knowledge assists the rulers to identify the common good as well as to mobilise the proper means to achieve it. People are then epistemically wronged, in the sense that their individual understanding of what is good, bad, desirable or undesirable might be uninformed or misinformed. If they do not behave in a way that contributes to the common good identified by the alliance of scientists engaged in political scientism and politicians, then they are ignorant. Thus, the judgements of people on how to self-determine their lives and their basic rights in that respect are overruled by the alliance of science-priests engaged in political scientism and rulers engaged in seeking to provide unlimited protection, that is, welfare totalitarianism.

4.5 The Illusion of Republicanism

Actually existing postmodernism in its totalitarian phase is a combination of several elements that were slowly creating the structures and incentives for it to emerge within society. A trend developed out of the combination of these elements that was rapidly gaining momentum and that was difficult to resist (see also Hadas 2021). Thus, the economist Philipp Bagus and co-authors (2021) show how welfare states can contribute to the formation of mass hysteria. The Flemish psychologist Mattias Desmet (2022, Part II) shows how a mass movement can then arise that leads to a totalitarian regime. The trend at issue is constituted by the following four elements:

- political scientism
- intellectual postmodernism
- welfare states
- crony capitalism

To use Marxist terms (appropriate here), political scientism and intellectual postmodernism are the intellectual superstructure, the *Überbau*, whereas welfare states and crony capitalism constitute the real-world substructure, the *Unterbau*.

At first glance, the corona regime may look like a panic reaction to a virus wave. Panic was certainly deliberately stirred up, especially by those who should actually keep a cool head and rely on the evidence, namely scientists, civil servants in leading positions and politicians. Nevertheless, deliberate spread of fear and panic is no explanation for what we have experienced since spring 2020. Panic does not last for several years.

It was striking that some of the medical experts who were portrayed in the media as the mouthpiece of science and who denigrated other, previously recognised professional colleagues, had already predicted a pandemic of a new magnitude in 2009–2010 with the swine flu. Back then, these scientists were stopped in time. Politicians and the media did not listen to them, but by and large kept a clear head and listened to scientists who warned against fearmongering, such as the epidemiologist Ulrich Keil (2010).

Medical scientists such as Anthony Fauci in the United States, Neil Ferguson in England and Christian Drosten in Germany presumably intended to use the next best virus outbreak to spread panic and call for a change from a medical to a political strategy with lockdowns and vaccination campaigns in the fight against the virus outbreak. The assumption that these scientists coordinated among themselves and with other like-minded people is certainly also plausible. But this is nothing new and nothing secret. It was known what these people wanted and that they were prone to fearmongering. It is just as well known what people like Bill Gates or Klaus Schwab want. They state this clearly in their publications (see the books Gates 2022 and Schwab and Malleret 2020, 2022). If one reads their outlets, one can deduce what kind of science these persons promote and strive to present as "The Science". There may be collusion among all these people. But this is not a conspiracy if only because it is no secret what these people seek to push through and by what means they intend to achieve their goals. If one nevertheless wants to speak of a conspiracy here, then one must simply acknowledge that such conspiracies always exist (see also the French *Manifeste conspirationniste* 2022).

A conspiracy is hence not an explanation for what we have experienced since spring 2020. This is not the result of a conspiracy by a world elite or powerful groups. When great evil happens, such as deliberately caused suffering and even deaths due to the lockdowns and the campaigns to treat people with insufficiently tested vaccines, then it is tempting to assume that the explanation is a conspiracy by a world elite or powerful groups who bring about this evil through their machinations—even if these people do not have evil intentions, but are convinced of their predictions of disaster and their proposals. However, such persons and the coordination among them always exist. The hypothesis of a conspiracy does not explain why the conspiracy is successful, that is, why so many people join in and do obviously absurd things. This cannot be explained in terms of a plan executed by an elite outside of—and opposed to—the general public.

The reference to the fact that the conspiracy goes hand in hand with profit interests on the part of the individuals involved in it and, above all, certain businesses, is no explanation either, although the corresponding connections exist for sure. Only a few people and businesses profit from the lockdowns and from the vaccination campaign. The theory of a conspiracy in combination with profit interests of certain corporations does not explain why so many people participate, that is, participate to their direct, obvious economic detriment. There are many more companies that were harmed by the lockdowns, testing, quarantines and vaccination requirements than there are companies that benefited from this regime. We have to explain why so many go along with this regime, against their economic interests and against their values and convictions in their past dealings with their fellow human beings.

The following fact also speaks against a conspiracy as an explanation: the same pattern of action that emerged in the reaction to the coronavirus waves also appears in the reaction to climate change and alleged oppressed minorities. As explained in Chap. 2, Sect. 2.7, the corona, the climate and the wokeness regime exhibit the same pattern. But the individuals, scientists and businesspeople who drive these regimes are in each case different, although there is some overlap and collusion between them.

To stress again, there always are scientists who dream of steering the courses of people's lives through "social engineering", there always are intellectuals that are prone to totalitarian ideas, fearmongers that hallucinate the end of the world and pseudo-philanthropists. There always are business interests that seek to profit from promoting these ideologies, as the pharmaceutical industry in the corona regime, the "renewable" energies industry in the climate regime and the ESG labelling in the wokeness regime. And there always is collusion between these groups. Such "conspiracies" do not explain the corona, the climate and the wokeness regime, because they are nothing new and distinctive of these regimes. What has to be explained, if one thinks that there are such "conspiracies", is why they are successful.

Consider the following thought experiment: imagine that Bill Gates, Klaus Schwab, the WHO director and whoever else one suspects of being involved in such "conspiracies" are forced to go into exile on an island where they are cut off from all communication with the rest of world. That would not change anything of significance. The mentioned four elements—political scientism, intellectual postmodernism, welfare states and crony capitalism—would persist. Actually existing postmodernism is a trend that results from the conjunction of these four elements and that manifests itself up to now in the corona, the climate and the wokeness regime.

Postmodernism as an ideology has slowly undermined the capacity of rational argumentation and evidence assessment in societies. Welfare states have undermined the rule of law and made individuals believe that they were entitled to numerous positive rights that necessitate the welfare state—as well as crony capitalism—for their fulfilment. Government-issued fiat money is the best system to finance the expenses of welfare states without directly and evidently taking ever more money from taxpayers. Scientism and political scientism emerge from modern science itself and provide the means to justify the overextension of welfare states. Postmodernism also gives them free runway as the limiting power of reason is dissolved by the postmodern ideology. The combination of these factors is what makes up the trend towards welfare totalitarianism and actually existing postmodernism.

In the first place, all four of these elements amount to an abuse of power. Science, through the technological advantages that it makes possible, is a power in the sense that one can do many things with its help. Through science we have achieved things today that people in earlier times could only dream of—like flying or communicating with other people anywhere in the world. Political scientism abuses the technological power that science confers for social engineering. Ideas are powerful in that they influence what people do. Intellectual postmodernism abuses the power of ideas in undermining the use of reason as a means to limit the exercise of power, both in science and society. Welfare states abuse the power that the authorities of constitutional states have through holding the monopoly of law making, law enforcement and judiciary by turning the state away from implementing the rule of law and towards fulfilling entitlements rights to various benefits. Crony capitalism then provides incentives for companies to abuse the power of the state authorities for their profit interests.

The hallmark of the use of power is not to make people do what they would do anyway out of their own free will and through employing their own reason. The hallmark of power is its use purely for the sake of exercising power. Thus, in the corona regime, in the name of science in combination with the other mentioned three elements, people were forced to do completely absurd things that they would never do of their own free accord—such as being persuaded that they are infected when they have a positive PCR test without having symptoms of a disease, wearing masks all the time in public spaces and thus harming their immune system, getting vaccinated out of "solidarity" without the virus being dangerous for oneself and even dragging one's children to a vaccination from which it was known from the outset that it cannot prevent the further spread of the virus, but that it may cause damage to one's health. In the climate regime, in the German *Energiewende*, all remaining safe and functioning nuclear power plants have been deliberately destroyed and coal plants reactivated for the sake of saving the world's climate from dangerous CO_2 emissions, although the nuclear power plants do not and the coal plants do emit considerable amounts of CO_2 (as well as really polluting substances).

Making people do completely absurd things is a test of the extent to which one holds power over them. Ultimately, however, power is not exercised for its own sake. The aim is to exercise power over people to create new human beings. As in all totalitarian regimes, so also in actually existing postmodernism, the aim is to transform the existing human beings into new human beings that comply with the ideal that the corresponding narrative sets—in the corona regime, human beings that no longer spread viruses (zero COVID-19); in the climate regime, human beings that no longer impact the environment through their use of energy (zero CO_2); in the wokeness regime, human beings that no longer express views that may hurt the feelings of somebody (zero tolerance). As in all past totalitarian regimes, so also in the present ones, the result, however, is only the destruction of the existing human beings by destroying both science and the rule of law.

However, each of the mentioned four elements was already present long before the postmodern totalitarian regimes emerged. Political scientism is as old as modern science; it can even be traced back to Plato, as mentioned in Chap. 2, Sect. 2.1. Welfare states and the crony capitalism that comes with them developed from the nineteenth century onwards. Intellectual postmodernism is with us for several decades by now. Hence, even if it is correct that the conjunction of these four elements explains the regimes of actually existing postmodernism, this explanation is not complete as it stands. We still have to answer the question why these four elements that have been with us for a long time combined only recently to constitute that trend.

Let us reconsider the fall of the Berlin Wall in November 1989 and the thesis of "the end of history" put forward by Francis Fukuyama (1989, 1992), meaning that a definitive victory of liberal democracies was achieved, at least in Europe and America. In the first place, to the extent that one is persuaded by this thesis, one is blind to the development that undermines liberal democracies through the combination of the mentioned four elements and that started already when "the end of history" was announced, although it became obvious only in spring 2020. The alliance among the forces driving these four elements arises from the middle of our societies. In contrast to the old enemies of the open society that Popper (1945) spotted in the totalitarian socialist regimes both of the communist and the national socialist type, the new enemies of the open society are situated not at its borders, but in the middle of our societies.

The new postmodern, collectivist narratives superficially tie in with the open society: they are globalist instead of nationalist, pluralist and inclusive—but not for the non-vaccinated, the fossil-fuels users, the meat eaters, the non-gender-language users etc., in short, all those who cling to their previous way of life and property rights, including the right of disposal over their bodies. In substance, pluralism and inclusion here do not go one iota further than the pluralism and inclusion of the then bloc parties in the Soviet empire. Globalism only holds for the elite who drives this regime. Everyone else is severely restricted in their opportunities for development and movement.

Nevertheless, the consequence of this linkage to the open society is that many friends of the open society fail to realise that we are on the way to a society that is closed under a new, specifically postmodern totalitarianism. Failure of awareness of what has become obvious with the corona regime since spring 2020 sadly is still widespread, also in the recent book by Fukuyama (2022) himself. People simply cling to the idea that what prevailed in 1989 was the open society and the principles of the republican rule of law as such. However, we have to take the idea seriously that this assumption is false. Under this assumption, one not only fails to understand what has happened in the meantime but is also misled into not even realising what is going on.

To put it differently, if what prevailed in 1989 had been the open society and the principles of the republican rule of law as such, the mentioned four elements would presumably still be with us individually. The thesis of "the end of history" does not signify the end of political scientism, intellectual postmodernism, welfare states or crony capitalism. Fukuyama himself takes liberal democracies to include a welfare state that promotes entitlement rights. However, these elements would not have combined to form the trend that fuels the new totalitarianism that we face.

Let us therefore reconsider the epoch between the end of the Second World War and the fall of the Berlin Wall. That epoch was characterised by the contrast between, on the one hand, the free Western societies and the communist bloc, on the other. For that reason alone, lockdowns would have been unthinkable when the Western societies faced the Asian flu of 1957–1958 and the Hong Kong flu of 1968–1970, although the coronavirus waves in 2020 were certainly not a greater health threat to the general population than these earlier severe virus waves. Of course, technological means were much more limited in 1957–1958 and 1968–1970 than in 2020. But that does not explain why no coercive political measures came into play at all in 1957–1958 and 1968–1970. No bans on major events were imposed then either. Economic and social life was maintained. Only sick people with evident symptoms stayed at home. The open societies and constitutional states of the West had to distinguish themselves from the communist regimes in Eastern Europe. The contrast between West- and East Berlin was obvious to everyone. Reacting to a viral wave with coercive political measures would not have been compatible with what the West stood for.

However, was this due to an appreciation of the open society as such? Or is the reason that society was held together by the distinction from communism and thus by a narrative that was specifically anti-communist, and it was incompatible with this narrative to react to a virus wave with coercive political measures? Answering the former question in the affirmative does not yield an explanation of the current development. It leads either to blindness with respect to that development or to the idea of a conspiracy against the people by a powerful elite. This, however, cannot explain the mentioned trend.

Answering the latter question in the affirmative yields this explanation: in accordance with what we set out as the West Berlin dilemma in Chap. 4, Sect. 4.3, the decades after the Second World War saw the development of a deep state in the West—both in the United States and in Europe—with a far-reaching security apparatus that intruded into the private sphere of people, industries that were tied to that state and a narrative that relied on demarcating the West from the communist bloc. The justification of that apparatus was evident, as in the first horn of the West Berlin dilemma: the real West Berlin would not have existed without the presence of the US military. When that narrative collapsed in 1989, the second horn of the West Berlin dilemma became manifest: the fall of the Berlin Wall notwithstanding, there still was the deep state. It was an illusion to think that this state apparatus would fade away once the threat from the Soviet empire had disappeared. Political power once acquired does not simply go away even when its rationale no longer obtains. When the narrative on which it is based collapses, what is to be expected is that it looks for another narrative to maintain and—if possible—extend its power.

To put it differently: the *Unterbau*—the welfare-warfare state with the crony capitalism attached to it—looks for a new *Überbau* to justify its power. That is the explanation of the alliance between on the one hand the welfare-warfare state and crony capitalism with political scientism and intellectual postmodernism on the other. As long as the anti-communist narrative prevailed, the state institutions were obviously opposed to postmodern and Marxist intellectuals and had no use for political scientism. However, once that narrative became obsolete, they had to look for a new narrative that—for lack of a powerful external enemy—could come only from within. We thus get an explanation of the willingness of the state institutions to receive intellectual postmodernism as well as political scientism. This enabled them to cling on to their power by staging new, metaphorical wars—against a virus, against climate change or against alleged social injustices. The supporters of the open society were unprepared and unarmed against this because they were under the illusion that the open society had prevailed qua open society. They did not consider that a narrative that depended on a relatively open society had become obsolete, so that it was to be expected that a new collectivist narrative would advance into this gap.

Hence, since it was not the open society qua open society that prevailed, but merely a narrative that was dependent on allowing a relatively open society for the cohesion of the society it serves, a gap then emerged in the form of the absence of a narrative. Into this gap a narrative was pushed that, while superficially tying in its rhetoric to the existing open society in order to conquer its institutions, in substance does what narratives tend to do when they are supposed to hold society together by an elite that sells its cling to power in terms of promoting the common good: establish a collectivism to which people must submit in their ways of life. It is only due to contingent historical circumstances that the narrative that prevailed until 1989 did not establish such a collectivism.

This explains why the institutions of the republican rule of law, up to and including the constitutional courts, fail; why so many people not only swim with the stream of the new collectivist narrative but also—and in particular the members of the leadership strata of what seemed to be an open society—promote this trend towards a new totalitarianism with full zeal; why the liberal and conservative parties in Europe and their supporters also join in etc. This is exactly what was to be expected when the narrative that contingently depended on the open society disappears, for this does not imply that the fact of a narrative that shapes and holds society together disappears. It is also to be expected that this process occurs slowly and takes time, roughly the time of a generation (i.e. roughly 30 years).

The mentioned illusion is what we propose to call in general terms "the illusion of republicanism". Its scope is much broader than yielding an explanation of the transition towards a new totalitarianism that we undergo. The illusion of republicanism is already evident in the two dilemmas into which the attempt to implement the rule of law in the guise of the modern constitutional state runs. Coming back to Hobbes' dilemma, the one horn is Hobbes' argument that in order to avoid a state of nature that is characterised by the uncontrolled use of violence because everybody enforces their own interpretation of natural law and the violation of natural rights, a sovereign has to be erected that holds the monopoly of law making, law enforcement and judiciary. The other horn of the dilemma is that if there is a state endowed with the power that this monopoly provides, the state organs are free to use this power as they see fit to protect the citizens up and including the declaration of a state of emergency in which basic rights are explicitly abrogated (West Berlin dilemma).

The illusion of republicanism is the belief that there can be a solution to these dilemmas by imposing controls from within the state that keep the state power checked without calling into question its monopoly of law making, law enforcement and judiciary. Historical circumstances can indeed be such that relatively stable republican states arise, as in the West in the decades between the end of the Second World War and the fall of the Berlin Wall. But the concentration of power in the hands of the organs of the republican state in the guise of the mentioned monopoly eventually leads, as any concentration of power does, to abuse if there are not particular historical circumstances that prevent such an abuse. It leads to the mentioned *Unterbau* that then requires an *Überbau* in the guise of a collectivist narrative and a legitimation that in the framework of modernity only science can provide.

Hence, science transgressing into scientism and from there into political scientism as explained in Chap. 2 is an internal development of science and this development is fuelled and promoted by state authorities intending to use science as a pretext to legitimate the extension of their power. Accordingly, restoring science requires going beyond the steps set out in Chap. 3. We also have to restore the rule of law. More precisely, however, if our diagnosis is correct, we cannot simply return to the rule of law as envisaged by the classical liberalism of the age of Enlightenment. We first have to overcome the illusion of republicanism. Austrian libertarianism provides the intellectual tools how to do so.

CHAPTER 5

How to Restore the Rule of Law

5.1 From Popper and Hayek to Mises and Rothbard: The Libertarian Turn

To overcome actually existing postmodernism, it is not sufficient to leave each of its regimes behind. The corona regime has by now largely faded away. However, the underlying scheme continues with the climate and the wokeness regimes. Even if these regimes were removed, it would have to be expected that actually existing postmodernism will fabricate yet other narratives to continue totalitarian social control. It is therefore indispensable to dismantle each of the four elements of the underlying trend described in the preceding section.

We already laid out the way how to restore science in Chap. 3 by cutting off the road from science to scientism: the very objectivity that is the hallmark of modern science entails that it can in principle not grasp human thought and action. Modern natural science therefore needs to be complemented by a human and social science that does not seek to apply natural science also in the domain of human thought and action in order to appear as scientific. As we argued, human freedom is the presupposition for doing science. Hence, such a human and social science has to acknowledge human freedom as established by a transcendental argument as its bedrock and work out how the trio of freedom, reason and normativity entails freedom rights. This, then, is the intellectual cure against scientism

M. Esfeld, C. Lopez, *Restoring Science and the Rule of Law*, Palgrave Studies in Austrian Economics, https://doi.org/10.1007/978-3-031-71186-2_5

and its political use. However, this is not sufficient. Science—both natural science and human science—can only blossom within a free society with a firmly established rule of law. Restoring science therefore requires also restoring the rule of law.

In his tribute to Friedrich von Hayek, Karl Popper (1997, p. 312) prides himself on having

> had some little influence in turning his [Hayek's] interest in this direction—simply because I stressed again and again in some of our conversations that the fight against what was then called, and criticized—especially by Ludwig von Mises—as protectionism, or state-protectionism, could not be conducted in the way of Mises; because in a complex society, anything approaching a free market could only exist if it enjoyed the protection of laws, and therefore of the state. Thus the term 'free market' should always be placed in inverted commas, since it was always bound, or limited, by a legal framework and made possible only by this framework.

Popper is entirely correct in pointing out that the free market "was always bound, or limited, by a legal framework and made possible only by this framework". The same applies to a free society in general, that is, a society based on the voluntary social interactions among its members. Such a society is only possible within the framework of the rule of law that enforces the obligation of every person to respect the freedom rights of every other person. Otherwise, the interactions in the market—and in society in general—would not be free interactions. However, the mistake lies in the conclusion from "the protection of laws" to "therefore of the state"—that is, to take the indispensability of the rule of law to entail the indispensability of the state. This means to endorse the first horn of Hobbes' dilemma (without the state, there would be lawlessness) without realising that one thereby runs into the second horn of Hobbes' dilemma (whatever the state sets as law is law within its territory). In fact, hence, Popper prides himself in this quotation on having persuaded Hayek to fall prey to the illusion of republicanism in his treatise *Law, Legislation and Liberty* (1982).

Hayek had therefore better not follow Popper's advice, but Mises' one. Indeed, Hayek (1952) and Popper (1957) are very clear about why natural science cannot be applied to human and social science. But they are less clear in setting out a positive and constructive alternative for a human and social science that does not follow the paradigm of natural science. Mises

(1949), by contrast, does so by elaborating on an a priori science of human action that he works out notably with respect to economics in the Austrian tradition. However, given that the trio of freedom, reason and normativity is the hallmark of human thought and action, this is an a priori science that is at the same time a normative science, as explained in Chap. 3, Sect. 3.4. The bedrock is not Mises' descriptive, a priori argument of consequences according to which free social interactions and thus a free market result in a more prosperous outcome for all participants than central planning and thus socialism (although this argument is entirely correct). The bedrock is the transcendental argument establishing human freedom and the normative argument of freedom rights that follows from this transcendental argument. That is why the normative turn from Mises (1949) to Rothbard (1982) is of central importance in establishing a positive and constructive view of human and social science in the Austrian tradition.

The normative turn is automatically a libertarian turn. Libertarianism in political philosophy follows from the transcendental, normative argument that establishes human freedom and freedom rights—although the transcendental argument is absent in Rothbard (1982) and brought into Austrian libertarianism by Hoppe (1989, chapter 7). The merit of Rothbard (1982) in this context is to have shown that the normative argument of freedom rights is at the same time a libertarian argument against the state. The reason simply is that it follows from this argument that coercion is legitimate and justified if and only if the person or group of persons against whom coercion is applied first did something—or set out to do something—through which they violate the freedom rights of other persons. However, states apply coercion indiscriminately and in a proactive manner by claiming the monopoly of law making, law enforcement and judiciary on a territory: everyone on the territory in question is coerced to obey the orders of the political authorities. Against the background of the theory of natural freedom rights, this coercion is illegitimate.

Kant works out the transcendental argument for human freedom and its consequence, namely the moral obligation to always recognise a person as an end in itself. However, he then fails to draw the conclusion that coercion is justified only if a person violated the natural rights of other persons by not respecting them as ends in themselves. By contrast, Kant goes then automatically from the rule of law to its implementation in a state without realising that the state authorities apply coercion indiscriminately (*Metaphysics of Morals* 1795, part 1, §§ 41, 42, 45). Popper thus is in venerable company when he takes the indispensability of the rule of law to entail the indispensability of the state. Nonetheless, this argument is fallacious as becomes evident when one enquires into the conditions under which coercion is legitimate.

More precisely, the step from the necessity of a stable legal order (rule of law) to the necessity of the state contradicts all the three arguments that limit the scope of this legal order summed up at the end of Chap. 4, Sect. 4.2:

- *The argument of epistemic modesty*: There is no knowledge concentrated in the authorities of a central state that could implement the rule of law through the monopoly of law making, law enforcement and judiciary in the hands of these authorities. The reason is that pluralism in trying out different strategies that compete with one another for the best way to enforce natural law always trumps a monopoly in the hand of a central authority. This is an instance of the general rule that the knowledge of the many in dealing with concrete problems always beats the knowledge concentrated in the hands of a few selected people when it comes to dealing with practical matters. This is an argument in favour of decentralisation and against the *Leviathan* as a central authority.
- *The normative argument of freedom rights*: Also the constitutional state violates the basic rights of persons against unwanted external interference in their life, limbs and property. The reason is the mentioned one that those persons on whom political authority is bestowed employ coercion against all the persons on their territory independently of whether or not these persons have done any acts through which they violate the freedom rights of other persons.
- *The argument of consequences*: The concentration of power in the hands of the state organs through their monopoly of law making, law enforcement and judiciary leads in the long run inevitably to a development that results in going down the road to serfdom, to put it in the terms of Hayek (1944): this monopoly has the consequence that the incentives for the holders of state power are such that the constitutional state declines into the welfare state with the road to welfare totalitarianism then open, finally leading to the various forms of collectivism and totalitarianism that we have experienced from the twentieth century on to this day, as explained in Chap. 4, Sects. 4.3, 4.4, and 4.5. The biggest crimes in history were not committed by

individuals or groups of individuals in stateless societies or dysfunctional states, but always by holders of state power in the name of the common good. In particular, the four elements whose conjunction drives the current regimes of actually existing postmodernism illustrate how easily also Western republican, constitutional states that look like implementing the rule of law and that succeeded reasonably well in doing so for a considerable period of time can go down the road to jettisoning the rule of law and ending up in totalitarian rule.

When the Enlightenment age set out to implement the rule of law and promote science through the creation of the modern central state, the philosophers and intellectuals in the liberal tradition overlooked the negative incentives that the concentration of power in the organs of the state would bring about. It was naïve to believe that the incentives that political power creates could be limited, without overstepping.

However, pointing out the illegitimacy of central states claiming the monopoly to interpret and to enforce natural law does as such not cut much ice. Hobbes concedes that no state in history has a legitimate origin. His argument for the state can be read in two ways. First, although no state has a legitimate origin, given that states exist as a matter of fact, it is better for the prosperity of human societies to keep the existing states than to dissolve them. This is an argument of consequences, as pointed out in Chap. 4, Sect. 4.3: dissolving the existing states would be detrimental for social relations since the conditions for mutual trust as a prerequisite for cooperation would then not be fulfilled (*Leviathan*, chapter 13). The second way is to lay out ideal conditions upon which the state could be justified. Consent-based theories seek to do this. Even though actual states are not justified (because they were not created out of consent), the state is not intrinsically pernicious because there would be conditions upon which it is justified (consent).

The entire liberal tradition—apart from libertarianism—endorses either of these arguments. From Locke on, the liberal tradition seeks either to come to terms with the existing states or to lay out some conditions for their justification (even when they were not met historically). The justification of the state in either of these forms is followed by arguments to limit the scope of the state through diverse structures such as the vertical division of power through a federal state structure as well as the horizontal separation of powers between a legislative, an executive and a judicial authority at all levels.

The libertarian case has then a twofold task. It not only has to provide cogent arguments against *any* justification of the state, even in hypothetical or ideal forms. Moreover—and more importantly—it also has to show that there are reliable mechanisms that can implement the rule of law in a stable manner without a state. Thus, libertarianism has to establish that the reasoning from the necessity of a legal order to the necessity of the state is unwarranted. We can have the former without the latter.

Against this background, it becomes clear why we cannot simply return to the path of modernity as it existed before actually existing postmodernism began its triumphant march. That is, we cannot simply pick up the thread from where we were before the development of always-growing welfare states—that is, from where we were, economically, before 15 August 1971, when the first, economic phase of actually existing postmodernism began with unleashed fiat money. And we cannot simply pick up the thread in society at the point where the Soviet empire collapsed in 1989.

If we cannot simply resume the previous path of modernity, then we have to at least recover its spirit. What is *needed is the dissolution of power in the hands of the authorities of the central state, just as at the beginning of modernity the power conglomerate of church and state was dissolved.* One can refer to the leitmotifs of the French Revolution in this regard: liberty, equality, fraternity. Also, for instance, Rudolf Steiner took up these leitmotifs in anthroposophy in such a way that he advocated a threefold structure of social life in which legal life (equality), economic life (fraternity), and intellectual life (liberty) are separated from each other (see the essays in Steiner 1966, 1985). Thus, there are various ways available to implement the leitmotifs of the French Revolution in a self-organising society based on voluntary interactions instead of a society organised through the concentration of power and coercion in the state organs.

In the first place, this means cutting off the link that leads from the constitutional state, seeking to implement the rule of law, to the welfare state. In a second—and even more difficult step—this means showing a way how there can be a stable rule of law without a state that holds a monopoly of law making, law enforcement and judiciary. Finally, if such a project is to succeed, we need an intellectual current of a new Enlightenment that replaces intellectual postmodernism. We go through all these steps in the following sections.

5.2 How the Open Society Can Flourish Without State Regulation

When Popper prides himself on having recommended to Hayek not to follow Mises, not only he has in mind that implementing the rule of law requires the state with its monopoly of force, but also that, moreover, the open society in the sense of Popper (1945) requires state regulation as the framework within which it can blossom. State regulation concerns not only the economy but also science, education, the media etc. As a matter of fact, also in the Western world since the period after the Second World War at the latest, not only the state authorities regulate these areas of social interactions, but governments intervene in the economy and run large parts of the institutions of science and education as well as the most influential media themselves, acquiring more or less a monopoly in offering these services and financing them through coercion (taxes and other compulsory levies).

Although Popper's (1945, 1957) diagnosis of the intellectual roots of totalitarianism is lucid and his advocacy of the open society exemplary, the argument for state regulation setting the framework within which the open society can flourish is simply fallacious. More precisely, it is fallacious by Popper's own standards. The reason is the argument that Popper (1957) and Hayek (1944, 1952) emphasise repeatedly, namely the argument of epistemic modesty: the knowledge of the many in dealing with concrete problems always beats the knowledge concentrated in the hands of a few selected people on whom political authority is bestowed and that hold a monopoly, thereby shedding competition. In short, the state is not the guarantor of the open society. Quite to the contrary, in virtue of the concentration of power in the hands of those who hold political authority, the state is the greatest threat to the open society. The reason is that this concentration of power inevitably leads to the development described in Chap. 4, Sects. 4.4 and 4.5. Hence, *restoring the rule of law requires in the first place to dismantle the power of the state organs. It requires removing their regulation of the voluntary social interactions of the people entirely.* For these interactions to flourish and thus for the open society to blossom, only the enforcement of the rule of law is needed in safeguarding the rights of defence of every person against unwanted external interference in their course of life. To put it in a nutshell, voluntary social interactions regulate themselves; only protection against involuntary interference is needed.

The first and foremost step in that direction is to abandon crony capitalism by rejecting state interference in the economy and by defending high levels of decentralisation, deregulation and liberalisation of economic activities. This would reorient economic incentives towards entrepreneurial activities instead of lobbying and rent-seeking. A free economy organises itself through voluntary interactions. Without interventions or partial-central planning of the economy, economic activities under the division of labour depend on what individuals prefer out of their own accord. Alternative forms of economic activity and voluntary cooperation are then likely to develop, such as worker-run cooperatives or private capitalist companies. They will emerge bottom-up instead of being imposed top-down. The self-organisation of the economy and the tendency to voluntarily cooperate with each other is not wishful thinking. People are not isolated individuals, but social beings who always organise themselves into communities based on mutual support and especially support for the weaker persons, even if these communities are only families and circles of friends. Every totalitarian system first sets out to destroy these communities. When the rulers face only isolated individuals, there are no longer any communities that can limit the state's penetration into all areas of life. In order for the state organs to expand their power, they must take over tasks that voluntary communities would otherwise fulfil. This applies in particular to the coverage of life risks, especially with regard to health and old age.

If health care were not subject to government regulation, nothing like the corona regime could be imposed. There would then be a pluralism of medical treatments and health insurances covering various risks that would compete with one another and that would have to be oriented towards the various preferences and risk assessments of their customers, or members of the respective cooperatives. Being able to make one's own decisions about one's health care and the desired type of medical treatment is an essential part of the freedom to shape one's life. The cartel organised by the state organs with compulsory levies and compulsory insurance would then be broken up. This would not only bring about a significant reduction in costs (as competence and competition would be the rule) but also impose further responsibilities and precautions to health providers. Nobody would then be able to force services on people, make immense profits out of it and remain unaccountable due to state protection. Medical practitioners could also not be threatened with revocation of their licences if they did not participate in this cartel and did not administer certain medicines or certain medical treatments, such as the corona vaccinations, to their patients. The provision of health services would then not be regulated by the state. Healthcare providers would in this way not only have to address the patients' needs and interests but also be more prudent and respectful, as they remain all the time accountable to their customers.

Although the harm that crony capitalism creates has become most evident in recent years with the health damages incurred by the corona regime, the most important step to remove crony capitalism and to achieve an economy based on voluntary social interactions is to abandon the state's monopoly on money and to ban fiat money. As explained in Chap. 4, Sect. 4.4, fiat money is the most important instrument of welfare states to expand their power and to go down the road to welfare totalitarianism. Fiat money is the economic basis of the regimes of actually existing postmodernism.

Restoring the rule of law therefore requires returning to private money through a free banking system. It is not enough just to replace fiat money again with the partial tying of money to a material asset, as under the Bretton Woods system until 1971. The root of the evil is the money monopoly in the hands of the coercive state power (see Hülsmann 2008). The state monopoly on money is the mechanism for the hubris of the constitutional state to usurp more and more tasks through the creation of money, thereby promoting special interests instead of equal rights for all and weakening the voluntary social communities. An economy based on free voluntary interactions (division of labour) requires money as a unit of account, a medium of exchange and a store of wealth. What is recognised as money is therefore established from below through the interactions of people. A state authority is not necessary for this. A legal system in the form of functioning jurisdiction and enforcement is sufficient. In the tradition of Austrian economics, Friedrich von Hayek (1976) sets out the case for free money. Most recently, Thorsten Polleit (2022a) shows a way out of the current regimes towards a free society based on free money.

Gold is the paradigm example of free money that cannot be multiplied through governments' printing presses. But free money does not have to consist in gold, silver or any other real good (commodity). A real good is anything that had (and still has) a non-monetary use before—and independently of—being used as money. Bitcoin (or another sort of freely created cryptocurrency) could also assume the role of free money. With cryptocurrencies, and bitcoin in particular, a very interesting new possibility emerges: something that is used as money in voluntary transactions without being a real good. Bitcoin is a simulation that refers to nothing outside itself and that takes the place of money as a real good. Blockchain technology is of course real. However, if, as in the case of bitcoin, the blockchain network does not refer to anything outside itself (i.e. is not employed to certify and transfer property rights in tokenised real goods), then it is a simulation that takes the place of real goods. Bitcoin simulates the properties of gold as the paradigm example of a free money by being scarce, divisible in small units and easily transferable and portable.

In that respect, bitcoin belongs to postmodernism. But it does not belong to the regimes of actually existing postmodernism. Bitcoin does not serve any use of power by a state authority, that is, through coercion. The use of bitcoin is based on a voluntary agreement. Bitcoin thus also bears an essential characteristic of modernity: limiting the exercise of power—in this case in the unlimited printing of money—through the use of reason.

Aspects of postmodernism could hence endure even if the regimes of actually existing postmodernism collapse. Postmodernism replaces reality with its simulation, and it replaces the use of reason to limit the exercise of power with the pure use of power. The former is the prerequisite for the latter. But the two are not equivalent. The question is this one: Do some people today switch to bitcoin only in the face of coercive state power, because switching to a simulation that is somewhat removed from state control is easier in some situations than switching to a real asset like gold? Or would a significant number of people prefer a simulation like bitcoin to a real asset like gold even if there were no coercive state power seeking to enforce a monopoly over money? One can argue that in voluntary interactions unimpeded by a coercive power, people would prefer the real good to the simulation (see Hülsmann 2023). They would choose real goods with suitable properties such as gold or silver as free money. Of course, they would use technological innovations such as a blockchain network to certify and transfer ownership rights to the real goods. However, thanks to bitcoin, this is an empirical hypothesis that is open to falsification if we return to free money.

In any case, in a free banking system, where the issuing of money is private, lending is limited. Banks do not have the possibility to create money out of nothing (or uncontrollably) and grant it as credit. They can only lend to the extent that they are covered by deposits at the bank that were previously saved by renouncing on consumption; at most, they can lend fiduciary money to a very limited extent in cases of high, solid reputation. Limiting lending in this way prevents the cycle of economic boom, which consists of nominal growth created by state fiat money that is not productivity growth and low interest rates. The boom is then followed by a bust, caused by money growth without productivity growth. Private money also ends the devaluation of labour against capital mentioned in Chap. 4, Sect. 4.4. It ends the current system that benefits the owners of capital at the expense of workers. Removing state control of the market through the fiat money system establishes the win-win situations that arise in the free market through voluntary cooperation.

The same goes for the entire intellectual sphere, including science, education and the media. Feyerabend is correct in criticising and going beyond Popper in that he advocates separating science and its institutions from the state, as religion and the church were separated from the state at the beginning of the modern epoch (Feyerabend 1975, chapter 19). It was clear that merging state and church just led to wars, persecution and brutal censorship. The same goes for merging science and state. This paves the way for political scientism taking over the institutions of the republican state.

To cut off the roots of political scientism, it is therefore necessary to decentralise and to de-regularise the production and transmission of scientific knowledge and education. The state should not intervene on how knowledge is produced, how it is transmitted and how it is funded. State officials have nothing to say about what is scientific knowledge and what it is not. However, by funding science, they necessarily have to take decisions on what counts as science and what does not do so. To mention just one current, controversial example: do gender studies belong to the scientific enterprise, or do they not belong to it? This is not for state officials to decide; but by allocating funding, they inevitably have to take decisions about the content of science.

As it is pernicious for an economy that the state intervenes in it by regulating prices, money supply or production, so it is equally pernicious for scientific knowledge that the state intervenes in its funding and thereby in its content (see Kealey 1996). As it is not true for the economic life of societies that to flourish state officials have to regulate supply and demand, so it is not true for science that the state's tutorship is required. The twentieth century provides us with ample evidence of how centrally planned societies lead to devastating failures and overwhelming misery and of why open societies flourish: they are able to incorporate the knowledge and decisions of the many over the few. Mutatis mutandis, scientific knowledge succeeds when freely acting individuals overtly discuss arguments and evidence. The genius of one or a few in power will never match the knowledge that many scientific communities can produce. Quite the opposite is the case. Political power is employed by the self-identified experts to use coercion to impose their view on the rest, hampering open-minded scientific enquiry.

For instance, the climate change narrative has been fostered by creating a new opportunity for academic career (before the 1970s, most climate scientists were physicists or meteorologists) and through pumping governmental money to climate-change-oriented projects (see Ball 2014). A legion of PhDs, postdoctoral researchers and young professors was thus growing during the 1980s and 1990s, flocking towards where the money was. Putting aside that states reallocate people's resources by compulsion, it is in this way that governmental money in sciences creates incentives that deter the free search for scientific knowledge: scientists' talent goes where state officials say that it should be according to a self-identified elite of experts.

State regulation of scientific knowledge and its technological exploitation is particularly harmful in the pharmaceutical industry (see Flanigan 2017). To begin, pharmaceutical regulations are morally wrong. They even cause many deaths. Most countries enforce pre-market approval requirements for pharmaceutical companies to sell a new drug. To meet pharmaceutical regulations not only is immensely costly but also takes a long time. Potentially lifesaving drugs are kept out of the market for years to satisfy all the necessary regulations, letting people die while they could have been saved if regulations were absent. Obviously, absence of state regulations does not mean absence of responsibility. Quite to the contrary, under the rule of law, every economic actor has to assume full civil liability for damages caused by the goods and services offered. That implies having to find insurance that covers the civil liability risks. Under the rule of law, neither there is a state that can exempt companies from assuming civil liability (as done in an immoral, if not outright criminal manner in the corona regime with the vaccines), nor is there a state that can prevent economic actors from offering their products and services in a free market for voluntary exchange. Furthermore, regulations also deter innovation: high costs and unbearable delays discourage scientists and the overall industry to try out new drugs, treatments or techniques. Once again, state intervention causes a structure of incentives that leads to hampering scientific debate, to deterring innovation, to harming people and, ultimately, to abandoning responsibility through persuading the holders of political power to waive civil liability.

The arguments that seek to justify various forms of state intervention upon the production and funding of science rely, in one way or another, on public paternalism. Without state tutorship, scientists will produce the wrong theories and people will adopt false beliefs, harming themselves. Without state regulations, private companies will put people's life at serious risks. People therefore need to be guided, protected and helped. As markets fail if not regulated, scientific and technological knowledge will fail if not intervened upon and regulated. Beside the moral objections to public paternalism, the basic assumption of these arguments is just wrong. As mentioned before, the knowledge of the few can never match the knowledge and creativity of the many when freely participating in an open society.

Even more importantly, as wealth is produced by people—and not by governments or central banks—scientific knowledge is also produced by people trained in the scientific methodology, and not by an elite of experts. It is an instrument for individuals to pursue their aims, as a collective epistemic and practical enterprise. Nobody invented the sciences or the scientific methodology. It was the result of countless individuals in their efforts to understand the world and to improve their lives. It was, and is, a trial-and-error process in which people have learnt which are the most suitable means for a certain end. Therefore, scientific knowledge is in no need of state tutorship, nor is it in need of regulations. Obviously, every learning process involves mistakes; but they are easier to overcome when everyone has a share and a say in the trial-and-error process. Central planners and regulators tend to recognise and correct error only when great harm has already been done. Competition is what leads to the confrontation with reality, whereas central planning and regulation avoid the confrontation with reality to the detriment of scientific and technological progress.

The state is not the guarantor of the freedom of science and the quality of education. Through its extensive monopoly of financing education and science through compulsory levies, it is rather their greatest threat. This threat becomes acute at the moment when the holders of state power pursue certain interests. Through the monopolistic funding of schools and universities, they can then enforce these interests. In order to overcome the ideology of actually existing postmodernism, it is therefore necessary to free the education system from state influence. This means that schools are established and financed by the respective voluntary social communities or private institutions. This is likely to imply that there is a plurality of schools with different pedagogical, ideological or religious orientations. A monopoly that enforces a particular orientation, such as political scientism, is then excluded.

However, the financial requirements of scientific research in the experimental sciences are considerable. It is unlikely that research that goes beyond purely theoretical work can be funded solely by voluntary social communities, tuition fees and donations. There are also obvious economic interests in research and the technical implementation of its outputs. The involvement of private companies in research and its funding is not a problem if there is no state authority that businesspeople can influence in order to impose their products on the people. If there is no such state authority, then companies have to convince customers of their products and they can be held liable for any damage caused. Research that is not oriented towards the interests of the people and that does not observe the method of critical examination of knowledge claims will then have short legs. If everyone has to assume liability, no one can afford to pursue ideology-driven research. Those who fund projects that do not respect the methodology of science would not find insurance to cover the resulting products against possible damages.

In general, free science and education can develop only if the influence, including the funding, of the state on science and education is removed. Like the economy, science and education can be based entirely on voluntary cooperation. Pluralism in education would thus be ensured through the various preferences that people—and in particular parents—have. Eliminating state-run scientific projects and state-funded educational and scientific institutions would open scientific production and reproduction to competition and more local incentives, allocating resources to the most urgent and needed projects, and not to where state agents centrally decide. This also amounts to reducing the size of the welfare state and one of its means to control societies.

An obvious objection to dismantling the concentration of power in the state organs is the following: If the economy, as well as education and science, are separated from state interference, is there not a danger that society will disintegrate into different, self-organising social communities? This objection is unfounded. These different, self-organising communities live together on the same territory under a legal order that enforces the obligation of all to respect the right of all others to live freely. This, however, is precisely the open society as outlined by Karl Popper (1945):

different ways of life, cultures, worldviews, religions etc., living together on one territory and mutually enriching each other through cultural exchange and the division of labour, respecting the human rights of all. *A state power that intervenes in the economy, education and science is the greatest danger to the open society instead of being its guarantor.*

The open society does not need a public space that is structured by state-funded media, educational institutions and science as well as a state-organised economy. Public space already forms itself when everyone can contribute their talents to an economy based on the division of labour. Different educational institutions then also compete with each other in the public space, which promotes their quality: no one voluntarily pays for what is conceived as poor-quality educational institutions. The same applies to a science that is ideologised instead of pursuing the quest for knowledge through disciplined scepticism. In short, exchange arises by itself through pluralism, provided that there is a legal system that consistently enforces equal rights for all.

5.3 How to Resist the Concentration of Power: The Rule of Law Without State Monopoly

It is clear that any free, open society based exclusive on voluntary cooperation requires a general, known rule that settles disputes and deters violation of rights, namely the rule of law. The decisive question is then the following one: Is it possible to guarantee and enforce the rule of law without a centralised state that claims the monopoly of law making, law enforcement and judiciary? In other words, the decisive issue is whether there is a way to overcome Hobbes' dilemma. To recall (see Chap. 4, Sect. 4.3), the one horn of the dilemma is Hobbes' argument that in order to avoid a state of nature that is characterised by the uncontrolled use of violence, a sovereign has to be erected that holds the monopoly of law making, law enforcement and judiciary. The other horn of the dilemma is that if there is a state endowed with the power that this monopoly provides, the state organs are free to use this power as they see fit.

There are indeed strong arguments developed in the vein of Austrian libertarianism that show a way out of this dilemma, namely, through the rule of law being realised in a decentralised manner without a state that enjoys a monopoly of law making, law enforcement and judiciary (see notably Leoni 1961; Rothbard 1982; Huemer 2013; Leeson 2014 and Dürr 2019). Let us highlight here three considerations in that respect.

(1) In the first place, the very idea of making law is misguided. Law cannot be made, by the same token as facts—scientific or otherwise—cannot be made (despite the original meaning of the Latin word "factum"). Natural law is discovered. Societies implement natural law through diverse public or private institutions, which might fail to implement it correctly. That is so when these institutions do not establish a firm rule in which each person is respected as an end in itself, which is the case today also in Western societies, as has become obvious at the latest with the corona, the climate and the wokeness regimes. The general point is that a highly centralised state is not the appropriate means to implement the rule of law because it paradoxically leads to its systematic violation.
(2) Furthermore, there is no need to settle in advance and for all possible situations what natural law demands. Also, there are good reasons to think that the implementation of the rule of law will differ across societies and cultures, taking into account various practices and traditions, despite its principles being the same. The crucial issue is to have the resources to settle disputes when they arise. The rule of law only demands that in the case of a dispute, any of the parties has the possibility to call in a neutral judge, and that the court's judgement be enforced through coercion (possibly subject to its validation through an instance of recourse). That is, it is sufficient for the rule of law to be implemented that (i) there is a judiciary that acts when called, that (ii) the judiciary is accessible to everyone to be called, and that (iii) there are institutions (public, communal, or private) that are in the position to enforce its judgements.
(3) It is crucial to bear in mind that since no one has a privileged access to the truth—there are no Platonic philosophers or other experts that have knowledge of the idea of the good—making room for pluralism and open debate is the way to discover what is right. This applies to science as well as to law. Judges only have to decide the concrete cases that are brought before them in the light of natural law and reason, instead of stating regulations that are supposed to cover all possible situations. Other judges decide the cases brought to them by taking into account the judgements of the earlier judges

> without being bound by them. There is no need for a central authority that appoints the judges either. That would be contrary to fostering pluralism in discovering what natural law demands in a concrete case.

Different communities may rely on alternative mechanisms to deal with disputes. Some may install a system of private protection and courts, while others develop communitarian systems. This, then, is the contrast between a private law society and one based on a codex of common law. Any such pluralistic system obviously is fallible. However, it is much less prone to error than a central state power exercising a monopoly of judiciary. That monopoly tends to abuse and to not pay attention to local circumstances and customs. Corrections are also much easier implemented in a pluralistic system in which several types of judges and courts compete for the just settlement of cases than in a monopolistic system.

Digging into the history of European law, Bruno Leoni (1961) shows how a stable legal order can be created based on judges finding out what natural law demands in concrete cases (see also Lottieri 2002). On this basis, then, common law emerges, develops and is corrected through a pluralistic judiciary system. That is to say, European common law provides a rich source how to establish a stable judiciary without a monopoly in the hand of central state authorities. In particular, Ireland has a long history of common law and a judicial system based on common law that was independent of the central state, and that existed in parallel—and in competition with—the British rule (e.g. see Sinder 2001; Jackson 2004; Laird 2005).

To consider another example, small villages in the Andes have for centuries developed a communitarian system of justice that was not only very effective to keep crimes at bay but also more responsive, quicker and cheaper than a centralised justice system.[1] Aymara communitarian justice sporadically emerges within the contemporary Bolivian state when ordinary justice fails or is out of reach. These communities were administrating justice for centuries according to rules that aimed to keep stability and cooperation in the community. They typically protect land and property rights (that belong in general to the family) and punish common crimes (such as thefts and murders). In communitarian justice, it is the *whole*

[1] See https://www.law.berkeley.edu/wp-content/uploads/2020/10/Indigenous-Bolivian-Community-Justice.pdf

community that exercises justice (even though the assessment of evidence and punishment is carried out by a justice assembly and elected authorities), and the whole community is responsible for justice (this is, for instance, evident in the "pacto de silencio"—"silence agreement"—that all members of the community keep after justice was administrated). The ordinary system of justice that the Bolivian state imposed after the Constitution of 1967 banned communitarian justice, making it illegitimate and punishing its exercise.

Conceptually, it is clear that a state, claiming the monopoly of coercion and *justice*, cannot allow alternative justice systems to run in the same territory (competition). Yet, the ordinary, monopolistic system of justice is out of reach for many communities (for instance, sometimes the closest police office is 100 km away and the transportation and time costs to denounce a petty crime are higher than a yearly salary) and the system is not effective in punishing crimes or corruption. This had the effect that corruption and crimes rose in Bolivia in the last decades and, consequently, that sporadic emergences of communitarian justice appeared in many communities and cities. The 2009 Constitution faced this issue and partially recognised communitarian justice as legitimate, although under the limits of the national constitution and the ordinary system of justice. Be that as it may, it is clear that demands for justice in some regions respond to local circumstances and cultures, and that the monopolisation of the administration of justice creates issues and negative externalities that can be avoided by a more pluralistic system (for Aymara communitarian justice, see Karp 1999; De Alarcón 2009; Craig 2009 and Hammond 2009).

Finally, participation and democracy do not have to be implemented in such a way that there is a tyranny of the majority over the minority. If the minority in a given vote is coerced to do what the majority has decided, this is coercion that easily leads to violations of natural rights (see the remarks on coercion in Chap. 4, Sect. 4.1). Participation and democracy also take place in voluntary social interactions without any coercion occurring when everyone can freely vote on the contracts in which they want to engage. That is, an open society implements participation through free markets, without any intervention into voluntary social interactions by state organs. Everyone can then vote on how to bring in their capacities in an economy of the division of labour and vote on which products and services to purchase given their means. In free market societies, customers, rather than producers, are the last judges and have control of an economy destiny. Anyone who offers goods and services has to win the votes of customers, without there being any possibility of coercing people by imposing something on them. Coercion only occurs when violations of natural rights take place and are recognised by the judiciary system. Once again, the crucial issue is the justification of coercion. Mere numbers of votes, as such, do not justify the application of physical force or the threat thereof.

A rule is by the very meaning of the word opposed to chaos, that is, the absence of rules. No society would function without rules. The rule of law is a regime to enforce natural law; coercion is legitimate as defence against or compensation for violations of natural law. In short, the *rule* of law can be conceived and implemented as a regime of rules without *rulers*, that is, without an authority that holds a monopoly on interpreting the rules (see Chartier and van Schoelandt 2021). Consequently, there is a way to avoid Hobbes' dilemma: the rule of law can be realised without a sovereign and thereby without the danger of sliding into totalitarian rule based on the monopoly of law making, law enforcement and judiciary that the sovereign exerts.

Furthermore, modern technology can be employed to strengthen this option: As mentioned in Chap. 4, Sect. 4.3, Hobbes' dilemma is a modern one. It results from the fact that modern societies are no longer organised along the lines of kinship and small communities based on personal acquaintance. The central political authority—the *Leviathan*—is intended as the solution to the problems of trust and coordination that arise because social interactions bring people together with whom one is not acquainted. Hobbes' dilemma then ensues because the central authority gives itself rise to a problem of trust: there is no reason to trust the holders of political authority that they will implement the rule of law impartially. The issue is not that politicians and state officials are morally bad people (although the power that comes with political offices and the opportunity to live comfortably at the expense of others attract people with a rather mean character). But the issue is the structural one of the incentives leading to the expansion of the power of those persons who hold political offices as explained in Chap. 4, Sect. 4.4.

Modern technology may indicate a way out of this dilemma. Consider an internet platform like eBay. Social interactions—exchange of goods and services—take place among strangers all over the world. Trust is created through a system of transparent evaluations of the transactions by all the parties engaged in each transaction. It is thus visible for all the participants who is a reliable transaction partner and who has a rather low score of reliability. A rather low score implies that one will have difficulties in finding transaction partners or that one can sell goods and services only at a lower price (internalising the costs of being a riskier seller). No central authority is required for that, just servers connected to the internet that register the evaluation of each individual party to a transaction.

In general, the vast majority of transactions worldwide occur smoothly without conflicts (negative reputation, for instance, is an always-in-force mechanism that deters scams or violations of contracts). Of course, there can be conflicts. However, again, no state is needed to settle them: the owners of the platform would just have to hire independent judges beforehand that hear and settle cases of dispute as well as hire an agency that enforces these judgements through the threat—and if need be the use—of coercion. Protection costs will be already included in the transaction fees, as happens today. Each of the participants in the transactions, by the very act of accepting the transaction, agrees to respecting a judicial procedure that is specified beforehand to settle potential conflicts. Hence, private platforms for social interactions have in principle the possibility to implement the rule of law for the transactions that they enable based exclusively on private law.

However, private platforms such as eBay are still central agencies, although they enjoy no monopoly like the state. Nonetheless, those who run the platform can in principle intervene at any time in any interactions that take place via the platform, cancel users, alter their scores etc. But since they have no monopoly, they compete with other platforms or services. Given that they cannot exert any coercion themselves, they would have to face the consequences of such interventions, which could lead to customers' distrust, bad reputation and the consequent loss of customers and thus revenue.

Moreover, there is a technological way out available: the use of blockchain technology is in principle in the position to replace central agencies, be they run by a state or by private companies. Blockchain technology enables certifying and transferring property rights in whatever goods—raw materials, precious metals, manufactured goods, real estate, works of arts etc.—efficiently and securely by tokenising these goods, either in the form of fungible or in the form of non-fungible—that is, unique—tokens.

No central agency that one has to trust is required for that. The blockchain in itself is trustworthy, because no single person or group of persons that use this technology can manipulate it. More precisely, the longer a blockchain is, the more difficult any attempt by an individual or a group of individuals becomes to manipulate it. Scale thus creates trustworthiness. Mutual evaluations of the participants in interactions via blockchain technology could in principle also be implemented via so-called smart contracts.[2]

In sum, the libertarian turn in the Austrian school follows from the normative turn. The libertarian turn does not deny that we humans are political beings. What distinguishes us humans from other social animals is the trio of freedom, reason and normativity. This makes it that our rule is the rule of law. The libertarian turn only consists in the claim that the step from the rule of law to a central authority on which the monopoly of law making, law enforcement and judiciary is bestowed—that is, the state—is a fallacy.

However, this does not mean removing the state and keeping only the economy in the guise of a free market. That is why the term "anarcho-capitalism" is a misleading characterisation of the libertarian turn: the issue is not capitalism, but voluntary social interactions, which can take many forms. The normative turn and the libertarian turn following from it do not imply a norm that stipulates that capitalist enterprises should be preferred to cooperatives. This is a matter of which types of economic interactions people prefer out of their own free will in self-determining their lives. The issue is not capitalism as such, but implementing the moral obligation to always respect every person as an end in itself in the guise of a stable rule of law without a political authority that exerts a monopoly on the interpretation and the enforcement of natural law.

Hence, this is indeed anarchy in the sense that there is no state that enjoys a monopoly of deciding when the use of force is legitimate. But it is not anarchy in the sense that there are no rules, as the slogan "anything goes"—Feyerabend's (1975) characterisation of anarchy—may be taken to suggest. Quite to the contrary, the issue is to firmly establish the rule of law. The libertarian argument is that the state that claims a monopoly of law making, law enforcement and judiciary fails in that respect. Libertarian political philosophy is about setting out a proposal how there can be a stable rule of law without a state in that sense. The bedrock is the principle

[2] We are grateful to Amine Rusi El Hassani for discussions on this subject.

of non-aggression as developed in Mises (1949). That is, the bedrock is a society based exclusively on voluntary social interactions, with coercion coming into play only as a reaction to a person or a group of persons violating the natural rights of other persons. The rejection of a monopoly of the use of force (the state), a free-market economy, free science and education etc. then all follow from this basic principle.

However, even if one can thus overcome Hobbes' dilemma as regards the internal organisation of a society based on the rule of law, the West Berlin dilemma remains. As explained in Chap. 4, Sect. 4.3, this dilemma concerns the situation of an internally free society that is surrounded by hostile powers. The dilemma then is that, on the one hand, an extensive military and secret service apparatus is required as defence against the external threat. On the other hand, the structure of this apparatus, notably the command structure of the military, compulsory military service and secret agencies that are not bound to respect fundamental rights is diametrically opposed to the principles of a free society and the legal order on which a free society is based and is poised to undermine this society.

There is no general solution to this dilemma. The prospect to cope with it depends on the concrete situation at hand. For instance, seeking to realise a society based exclusively on voluntary social interactions in West Berlin between 1945 and 1989 would not have been a good idea. Such an attempt would have immediately been crushed by the Soviet tanks stationed in East Berlin. As mentioned in Chap. 4, Sect. 4.3, mere non-interference with the affairs of foreign powers is no way out of the West Berlin dilemma; the mere existence of a free society in their vicinity destabilises their system of rule. Hong Kong taken over by Chinese rule (despite an international treaty with the United Kingdom to the contrary) and today's situation of Taiwan are contemporary examples of the West Berlin dilemma. Also Rothbard (1982, part V, chapter 30) concedes that the possibility of achieving a society based exclusively on voluntary social interactions depends on contingent external circumstances.

Again, new technology may be of help: the operation of a central platform for voluntary exchange such as eBay depends on the accessibility of certain servers at certain locations; governments can block the access to these servers. However, it is much more difficult, if not outright impossible for governments to prevent the people under their control from using blockchain technology. Bitcoin is a case at hand. Even if governments were to shut down access to the internet on a certain territory, people could still use radio transmitters or other technology for transactions implemented on a blockchain. Again, this is evidence of the two-sidedness of technological progress: it creates new means for governments to increase surveillance. But it gives people also the means to escape from government surveillance and thereby undermine totalitarian rule. In short, realising a society that is based exclusively on voluntary social interactions not only is an internal, local affair but may also depend on the worldwide availability of technologies that make central authorities as solution to the problems of trust and coordination superfluous.

5.4 How to Return to Reality Through a New Enlightenment

Restoring the rule of law requires not only dismantling the concentration of power in the hands of the holders of political authority but also inspiring an intellectual movement of a new Enlightenment that replaces intellectual postmodernism. However, in doing so, one must refrain from seeking to re-educate people such that they become somewhat more virtuous human beings. Obviously, any programme of centrally organised education or re-education would contradict the rule of law and the principles of a society based on voluntary interactions.

As we said in the introduction, people have various beliefs, interests, values, capacities and virtues. There will always be intellectuals that are prone to impose their views and values upon others, fearmongers predicting apocalypses and scientists that misconceive science as a secular religion. There will also always be business interests that seek to profit from promoting these ideologies. And, as Adam Smith (1976 [1776]) already warned us, there will always be collusion between these groups against others. The first and foremost step to avoid any totalitarian mindset is to recognise people's creativity and variety as well as their values as ends in themselves and never as a mere means. A free society can only flourish when collectivist narratives are abandoned, when compulsory ideals of virtues are rejected and when state coercion is diminished.

The corona regime, the climate regime and the wokeness regime would be doomed from the outset if they could not rely on the coercive organs of the modern states to impose their totalitarian rules upon the people. The same goes for totalitarian plans like the "pandemic preparedness" of the WHO, or the "great reset" (Schwab and Malleret 2020; see Rectenwald 2023). As we argued in this book, once states are endowed with the concentration of power given by the monopolies of law making, law enforcement and judiciary, their organs tend to maintain and extend their power and thus tend to fall prey to such totalitarian ideologies. These are always latent and activate themselves under particular historical circumstances, such as the current ones as explained in Chap. 4, Sect. 4.5. It is therefore hopeless to believe that the coercive power of the states can eradicate the disasters that coercive state power allows and boosts; for that very power is the root—and not the cure—of the problem.

Keeping political power at bay is nonetheless only one aspect of the solution. It is also necessary to face reality and recover our own judgement by removing the distorting ideologies. Restoring science and the rule of law crucially depends on removing both the *Unterbau* of actually existing postmodernism—the welfare state and the crony capitalism that comes with it—as well as its *Überbau*, namely the narratives fed by political scientism and intellectual postmodernism. Removing these four elements will enable us to return to reality and to leave behind the postfactual constructions and simulacra of actually existing postmodernism, both as regards the *Unterbau* with the fiat money as the postfactual simulation of wealth and as regards the *Überbau* with the narratives that impose a postfactual reality on us based on simulations.

In virtue of their monopoly of law making, law enforcement and judiciary, state authorities tend to use coercion to impose their particular views of society upon the people: a state's utopia then becomes a totalitarian utopia. However, when coercion is limited only to the protection of everyone's rights, then there is no single utopia, but many. As Robert Nozick explains in the third part of his book *Anarchy, State, and Utopia* (1974), the rule of law sets a framework for utopia in which each person is free to pursue their utopia and to do so through voluntary association with other persons. A free scientific system will also set a framework in which different theories and technological developments can compete, be publicly assessed and pursued. Voluntary social interactions not only overperform what states can do but also orient scientific knowledge, talent, creativity and resources towards where they are needed most. Any theory and technology will be assessed according to how it serves others, how it improves their lives and how it fares with respect to evidence and argument.

This framework enables us to escape the ongoing narratives (and future ones). The corona regime does not only exemplify totalitarianism by the imposition of draconian measures, but it also impoverishes the worse off, costs lives and resources and enormously contributes to social discomfort.

The corona regime is so costly because it is totalitarian and deeply unscientific; because it impedes the trial-and-error process whereby individuals (and societies in consequence) learn and succeed. Draconian and groundless measures were sold as the only possible solutions, justified by "The Science" and its alleged moral superiority. Dissensus was thus oppressed, and with it, the free exercise of judgement and the scientific methodology. But, without dissensus, societies cannot try, cannot learn, and science as disciplined scepticism cannot be exercised. However, dissenters were accused of immorality and unscientific reasoning. Many of their proposals were not even heard and examined. For instance, the Great Barrington Declaration (Bhattacharya et al. 2020) was denigrated in mass-media and by government-funded "experts" for pursuing a "let it rip" approach, when the main message of the proposal was a focused approach that protects those at most risk without harming the rest of society (first and foremost, the worse off). The Great Barrington Declaration approached the situation from the right, scientific angle: societies always face a trade-off in which *all* variables should be properly assessed by different individuals in the society according to their preferences, values and risk evaluations. A free scientific system impedes that the knowledge of a few experts prevails over the knowledge of the many. It also impedes that the mistakes and the conceit of the few be paid for by everyone.

The climate change narrative can also be escaped from by a framework for science in which states do not intervene and private technological innovation and individual assessments have a greater role to play. The first step to escape the narrative is to recognise the intrinsic trade-off that exists, and it is frequently hidden, at the bottom of the discussion. Economic growth and prosperity in developed countries was fuelled by secure, reliable and cheap energy (such as fossil fuels). It was in turn technological innovation that made secure, reliable and cheap energy available, depolluting households and drastically lowering not only fire-domestic deaths (e.g. with the transition from coal to natural gas) but also the death toll due to unfavourable weather conditions in general (see Chap. 2, Sect. 2.5 and Goklany 2011). Hence, "going green" is not cost free, but it implies slower growth (or even degrowth). Individuals have the right to assess this trade-off and the pros and cons of "going green", according to their circumstances and preferences.

The second step is to emphasise the importance of technological innovation to improve energy production and transportation. Companies use fossil fuel energies not because they are greedy and do not care about the future (entrepreneurial activities in general care more about the future than the present), but because it is simply better. By the same token, companies will stop using fossil fuel energies when a viable alternative is at least equally secure, reliable and affordable. If there is any climate problem to deal with, the best strategy is to completely de-regulate technological innovation in the energy sector and to incentive private investment (through, for instance, tax cuts). The best framework to do this is one in which technological companies and scientists are free to try out different strategies and ideas. Energy is a highly profitable sector. Capital will flock where the best and most promising ideas are being developed. It is unnecessary that states pomp tons of taxpayers' money into unpromising paths (such as so-called renewable, green energies) or that they regulate technological investments. Once again, the solution to any societal problem will come from the individuals within the society and not from the central planners.

Finally, a free scientific system is the best framework not only for people (experts and non-experts) to judge evidence and arguments but also for funding to move bidirectionally. Both elements just make scientific knowledge better by avoiding the creation of made-up consensuses. It puts us in a better position to assess risks and trade-offs, to form judgements according to our values and information, to weight different viewpoints and to make our own decisions on a more solid basis. Good theories will impose themselves because they can be overtly tested, criticised and developed. If good, they will make us better off and individuals will have incentives to adopt them if they adjust to their aims. Bad theories will just agonise and be discarded. But the only way to distinguish between them is not by referring to an elite of experts, but by their development in a free scientific system where the evidence and the arguments can be assessed by everyone according to well-known rules.

No more is necessary than to remove the narratives. The development of freedom cannot be steered, apart from assuring the existence of a judiciary system that enforces liability of the actions of everyone, thereby granting equal rights to all. If the state power is dismantled step by step, the economy including social security systems, science and education will develop out of their own, given in particular the enormous economic means that will be set free when the state apparatus is diminished.

When the intervention of the state organs upon society and science is removed, the proponents of the narratives will not find the means to coercing others to endorse them. It would be a mistake to seek to build and unify the open society on a narrative of freedom and self-determination that takes the place of the postmodern and collectivist narratives. Free, open societies do not—and cannot—depend on making people more virtuous. The issue is to remove coercion and to unleash the non-coercive incentives that foster voluntary interactions and cooperation. Different narratives and moral systems are welcome to the extent that they agree on the categorical imperative to always respect each person as an end in itself and never use a person as a mere means to achieve one's own ends. There is no need to streamline or unite such different narratives into one. The motivation for respecting this categorical imperative may stem from religion as in the Christian religion, from humanitarian considerations, from universal reason as in the transcendental argument set out in Chap. 3, Sect. 3.4, or simply from the consideration of the beneficiary consequences of voluntary cooperation (recall the normativity argument and the argument of consequences summed up at the end of Chap. 4, Sect. 4.2). The rule of law is a framework within which various utopias can blossom, that is, various ways in which individuals and voluntary communities fathom their lives according to their conceptions of what a good life consists in (see again Nozick 1974, part III). No dominant narrative that unites society is required.

This, again, is the tightrope that the use of reason as a means to limit the exercise of power requires us to walk (see the end of Chap. 3, Sect. 3.4). There is moral knowledge in the guise of knowledge of the categorical moral obligation to always treat every person as an end in itself and never as a mere means to an end. But this is only a formal knowledge of the freedom that each person enjoys in thinking and acting and thus only a normative knowledge of the moral obligation to respect this freedom. It cannot be a knowledge of a normative content of goals, values, or meaning for the life of a person or a common good for a society.

Hence, realising a society based on voluntary interactions requires on the one hand implementing this moral obligation through the rule of law and the application of coercion when violations of this obligation occur. On the other hand, this requires abstaining from imposing particular moral systems upon others. Particular moral systems go far beyond the moral obligation to always respect the negative freedom rights of defence of every person by, for instance, imposing various obligations to aid others, and they may well be justified on religious or humanitarian grounds. But they cannot be imposed on others through the threat—and eventually the use—of coercion. That is why acting in accordance with them is praiseworthy. By contrast, respecting the requirements of the rule of law is not praiseworthy as such. This is simply what one is obliged to do in virtue of having the normative status of a person. In short, however praiseworthy humanitarian actions particular moral systems may recommend, it would be immoral to seek to impose such systems through coercion. One can only convince others to join in the realisation of what one considers to be a (or the) common good by argument and setting an example in one's actions, but never by coercion.

This is also the fallacy of democracy: holding majority votes on various proposals for common goods does not make it legitimate to impose the realisation of such goods through coercion as in the welfare state; doing so obviously violates the unconditional freedom rights of persons. It amounts to treating those people who have not voted in favour of the proposal at hand as mere means to the realisation of what other people consider to be a common good, however big the number of these other people may be. Granting basic rights of defence against unwanted external interference in one's way of fathoming one's life is not a matter of majority vote. Hence, majority votes come into play only insofar as imposing what the majority decides upon those who did not vote in favour of the proposal in question does not interfere with their rights to self-determine their lives.

In sum, removing the narratives and overcoming intellectual postmodernism requires intellectually a new Enlightenment that emphasises the distinction between on the one hand the categorical moral obligation to always treat each person as an end in itself (rule of law) and on the other hand various moral systems and their visions of the common good—the various utopias in the sense of Nozick (1974, part 3) that each can be realised by the groups that adhere to them within a society based on voluntary interactions.

The *first step* towards realising such a society and removing actually existing postmodernism is to use one's own judgement instead of letting self-proclaimed experts take away one's judgement. That alone will dispel the grip of the postmodern narratives on the people and thereby lead to a limitation of the state's power over the people. The *second step* is to exercise the courage of making free public use of one's reason and to articulate scepticism about the concentration of power. This, then, is the basis to remove the state monopoly of coercive power and to build society on voluntary interactions within a legal system that implements the rule of law as the *third step*.

According to Kant's famous essay answering the question "What is Enlightenment?" from 1784 (English translation in Kant 1996, pp. 11–22), the courage to make free public use of reason is the essential step towards enlightenment. The dissolution of power over people in the hands of a state authority brought about by enlightenment requires the courage to trust one's own judgement and to express it publicly. In contrast to military courage under the command of a powerful warlord, civil courage can only develop from below and anarchically, without domination and of one's own accord. Everyone can contribute to this.

The regimes of actually existing postmodernism shy away from the open display of physical force in our societies. These regimes thrive on indoctrination, on people allowing their judgement to be taken away and following the supposedly scientific and supposedly moral guidelines that these regimes set and that replace the confrontation with reality with powerful postfactual simulations. There are no Soviet tanks today to sustain a totalitarian regime, even if no one believes in the narrative anymore. When enough people refuse to "Follow the Science", the regimes of actually existing postmodernism are finished. This is the chance that we have to make public use of our own reason to stop these regimes before the consequences of their ignoring reality show up in economic and social collapse. If enough people muster this courage, then modernity has a chance again with the use of reason as a means to limit the exercise of power.

When Kant answered the question "What is Enlightenment?" in his essay in 1784, he thought that he was living not in an enlightened age, but in an age of progress towards enlightenment. Our age of actually existing postmodernism, by contrast, is an age of regress away from enlightenment through the dazzles of postmodernism that cloud the use of reason. We therefore need a new Enlightenment. But it has to be an enlightenment that avoids the mistake of the Enlightenment of the eighteenth century, namely, to assume that there can be an enlightened central state promoting science and implementing the rule of law in a stable way, without sooner or later going down the road towards totalitarian rule. The new Enlightenment has to leave behind Hobbes' dilemma of there being either isolated individuals resorting to uncontrolled violence or a state that exercises the monopoly of law making, law enforcement and judiciary by showing how voluntary social interactions and cooperation can provide the protection for which the state is allegedly inevitable, but through which it slides from the constitutional into the totalitarian state. That notwithstanding, the new Enlightenment can and should rely on the way and the means that Kant set out to achieve enlightenment: the free public use of reason. If enough people have the courage to do so, this will be sufficient to realise the framework of the rule of law within which various utopias can blossom.

To put it in a nutshell, actually existing postmodernism replaces the contact with reality with simulacra, both as regards the substructure (notably fiat money) as well as regards the superstructure with the narratives that lay the ground for the totalitarian corona, climate and wokeness regimes. Overcoming these regimes requires returning to reality through a science that is free in the pursuit of truth and a society that is based on voluntary interactions. To obtain this, we need a new Enlightenment that resumes the path of freedom and progress by avoiding the illusion of republicanism to which the eighteenth-century Enlightenment fell prey.

Through science and the rule of law, our civilisation realised a previously unimaginable technological and economic progress that stretches out to all segments of the population. Based on the use of universal reason, our civilisation achieved the implementation of equal rights for all and abolished slavery. Continuing on this path, we can accomplish the rule of law and remove the monopoly of the state. It only requires the courage to make free public use of our reason and the courage to let voluntary social interactions blossom.

Bibliography

Ärtzeblatt (2020): "RKI-Präsident Wieler: 'Entwicklung in Deutschland macht mir große Sorgen'", https://www.aerzteblatt.de/nachrichten/115057/RKI-Praesident-Wieler-Entwicklung-in-Deutschland-macht-mir-grosse-Sorgen, 28 July 2020.

Afriat, Alexander, Esfeld, Michael and Cullen, Paul (2023): "Can the definitions of SARS-Cov-2 and Covid-19 stand up to epistemological scrutiny?", *Qeios*, 6 October 2023, DOI https://doi.org/10.32388/VCFIQW, https://www.qeios.com/read/VCFIQW

American Association for the Advancement of Science (2019): "How we respond. Community responses to climate change". https://howwerespond.aaas.org/report/

Amlinger, Carolin and Nachtwey, Oliver (2022): *Gekränkte Freiheit. Aspekte des libertären Autoritarismus.* Berlin: Suhrkamp.

Armstrong, David M. (1968): *A materialist theory of the mind.* London: Routledge.

Axfors, Cathrine and Ioannidis, John P. A. (2021): "Infection fatality rate of COVID-19 in community-dwelling populations with emphasis on the elderly: an overview". *medRxiv preprint*, DOI https://doi.org/10.1101/2021.07.08.21260210.

Bader, Ralf M. (2017): "Counterfactual justifications of the state". *Oxford Studies in Political Philosophy* 3, pp. 101–131.

Bagchi, Sutirtha, Curran, Michael and Fagerstrom, Matthew J. (2019): "Monetary growth and wealth inequality". *Economics Letters* 182, pp. 23–25.

M. Esfeld, C. Lopez, *Restoring Science and the Rule of Law*, Palgrave Studies in Austrian Economics,
https://doi.org/10.1007/978-3-031-71186-2

Bagus, Philipp (2010): *The tragedy of the Euro*. Auburn, AL: Ludwig von Mises Institute.

Bagus, Philipp, Pena-Ramos, José Antonio and Sanchez-Bayon, Antonio (2021): "COVID-19 and the political economy of mass hysteria". *International Journal of Environmental Research and Public Health* 18, 1376.

Bagus, Philipp, Pena-Ramos, José Antonio and Sanchez-Bayon, Antonio (2022): "Capitalism, COVID-19 and lockdowns". *Business Ethics, the Environment & Responsibility* 32, pp. 41–51.

Ball, Tim (2014): *The deliberate corruption of climate science*. Apache Junction, AZ: Stairway Press.

Baltrusaitis, Justinas (2020): "Working hours to buy S&P 500 sky-rocket to all-time highs in 160 years". *Finbold News*, 25 May 2020, https://finbold.com/working-hours-to-buy-sp-500-skyrocket-to-all-time-highs/

Beck, Konstantin and Widmer, Werner (2020): "Corona in der Schweiz. Plädoyer für eine evidenzbasierte Pandemie-Politik". https://www.corona-in-der-schweiz.ch

Beck, Konstantin, Kley, Andreas, Rohner, Peter and Vernazza, Pietro (eds.) (2022): *Der Corona-Elefant. Vielfältige Perspektiven für einen konstruktiven Dialog*. Zürich: Versus.

Bendavid, Eran et al. (2021): "Assessing mandatory stay-at-home and business closure effects on the spread of COVID-19". *European Journal of Clinical Investigation* 51, e13484.

Bendavid, Eran and Patel, Chirag J. (2024): "Epidemic outcomes following government responses to COVID-19: insights from nearly 100000 models". *Science Advances* 10, eadn0671.

Beran, Harry (1987): *The consent theory of political obligation*. London: Routledge.

Berlin, Isiah (1969): "Two concepts of liberty". In I. Berlin: *Four essays on liberty*. Oxford: Oxford University Press. Pp. 121–154, 169–172.

Bhattacharya, Jay, Gupta, Sunetra and Kulldorff, Martin (2020): *Great Barrington Declaration*, 4 October 2020, https://gbdeclaration.org/

Bjørnskov, Christian (2021): "Did lockdown work? An economist's cross-country comparison". *CESifo Economic Studies*, 29 March 2021, 1–14, DOI: https://doi.org/10.1093/cesifo/ifab003

Block, Walter (1986): *The US bishops and their critics*. Vancouver: Fraser Institute.

Block, Walter (2019): *Property rights: the argument for privatization*. New York, NY: Palgrave Macmillan.

Boudry, Maarten and Pigliucci, Massimo (eds.) (2018): Science unlimited? The challenges of scientism. Chicago, IL: University of Chicago Press.

Bradley, Raymond S. (2015): *Paleoclimatology. Reconstructing climates of the quaternary*. Amsterdam: Elsevier. Third edition.

Brandom, Robert B. (1994): *Making it explicit. Reasoning, representing, and discursive commitment.* Cambridge (Massachusetts): Harvard University Press.

Bundesverfassungsgericht (2022): "Erfolglose Verfassungsbeschwerde gegen die Pflicht zum Nachweis einer Impfung gegen COVID-19 (so genannte „einrichtungs- und unternehmensbezogene Nachweispflicht")". *Beschluss vom 27. April 2022*, Pressemitteilung Nr. 42/2022, 19 May 2022.

Burrows, Edwin G. and Wallace, Mike (1998): *Gotham: a history of New York City to 1898.* Oxford: Oxford University Press.

Caplan, Bryan (2021): "Loyalty oaths compared: an Orwellian exercise". https://www.johnlockeinstitute.com/post/loyalty-oaths-compared-an-orwellian-exercise, 26 February 2021.

Chalmers, David J. (1996): *The conscious mind.* Oxford: Oxford University Press.

Chartier, Gary and van Schoelandt, Chad (2021): *The Routledge Handbook of anarchy and anarchist thought.* London: Routledge.

Childs, Roy A. (1977): "The invisible hand strikes back". *Journal of Libertarian Studies* 1, pp. 23–33.

Chin, Vincent et al. (2021): "Effect estimates of COVID-19 non-pharmaceutical interventions are non-robust and highly model-dependent". *Journal of Clinical Epidemiology* 136, pp. 96–132.

Christy, John R., Norris, William B., Spencer, Roy W. and Hnilo, Justin J. (2007): "Tropospheric temperature change since 1979 from tropical radiosonde and satellite measurements". *Journal of Geophysical Research* 112, D60102.

Connolly, Roman et al. (2021): "How much has the Sun influenced Northern Hemisphere temperature trends? An ongoing debate". *Research in Astronomy and Astrophysics* 21.6, pp. 131-1–131-68.

Craig, Nathan (2009): "Modern Aymara justice systems considered in light of modern evolutionary theory". Manuscript. https://www.academia.edu/211409/Modern_Aymara_justice_systems_considered_in_light_of_modern_evolutionary_theory

Dagsvik, John K., Fortuna, Mariachiara and Moen, Sigmund Hov (2020): "How does the temperature vary over time? Evidence on the stationary and fractal nature of temperature fluctuations". *Journal of the Royal Statistical Society. Series A* 183, pp. 883–908.

Dagsvik, John K. and Moen, Sigmund Hov (2023): "To what extent are temperature levels changing due to greenhouse gas emissions?". *Statistics Norway. Discussion Papers* No. 1007, September 2023.

Davies, Stephen (2004): "The great horse-manure crisis of 1894". *Foundation for Economic Education* (FEE), https://fee.org/articles/the-great-horse-manure-crisis-of-1894/, 1 September 2004.

De Alarcón, Silvya (2009): "La justicia comunitaria como sustento de la ética Aymara". *Revista Integra Educativa* 2, pp. 253–278.

Deleuze, Gilles (1962): *Nietzsche et la philosophie*. Paris: Presses universitaires de France. English translation *Nietzsche and the philosophy. Translated by Hugh Tomlinson*. London: Athlone Press 1983.

Delfino, Victor (1926): "Eugénica". *Medicina Argentina* 51.

Descartes, René (1902): *Œuvres de Descartes. Tome 6. Publié par Charles Adam et Paul Tannery*. Paris: Cerf.

Desmet, Mattias (2022): *De psychologie van totalitarisme*. Kapellen: Pelckmans. English translation *The psychology of totalitarianism. Translated by Els Vanbrabant*. London: Chelsea Green Publishing.

Doshi, Peter (2010): "Rapid response: More changing web-pages at WHO". *British Medical Journal* 340, c1904.

Dürr, David (2019): "The inescapability of law, and of Mises, Rothbard, and of Hoppe". *Journal of Libertarian Studies* 23, pp. 161–170.

Duisenberg, Willem F. (2002): "International Charlemagne Prize of Aachen for 2002. Acceptance speech by Dr. Willem F. Duisenberg, President of the European Central Bank, Aachen, 9 May 2002." https://www.ecb.europa.eu/press/key/date/2002/html/sp020509.en.html

Dworkin, Gerald (1972): "Paternalism". *The Monist* 56, pp. 64–84.

Dworkin, Gerald (2013): "Defining Paternalism". In: C. Coons and M. Weber: *Paternalism. Theory and practice*. Cambridge: Cambridge University Press. Pp. 25–38.

Esfeld, Michael (1995): *Mechanismus und Subjektivität in der Philosophie von Thomas Hobbes*. Stuttgart- Bad Cannstatt: Frommann-Holzboog.

Esfeld, Michael (2020): *Science and human freedom*. London: Palgrave Macmillan.

Feinberg, Joel (1973): *Social philosophy*. Englewood Cliffs, NJ: Prentice-Hall.

Ferguson, Neil M. et al. (2020): "Impact of non-pharmaceutical interventions (NPIs) to reduce COVID-19 mortality and healthcare demand". *Imperial College COVID-19 Response Team*, 16 March 2020, DOI https://doi.org/10.25561/77482

Feyerabend, Paul (1975): *Against method*. London: New Left Books.

Feynman, Richard P. (1955): "The value of science". *Engineering and Science* 19, pp. 13–15.

Flanigan, Jessica (2017): *Pharmaceutical freedom*. Oxford: Oxford University Press.

Fodor, Jerry A. (1974): "Special sciences (or: The disunity of science as a working hypothesis)". *Synthese* 28, pp. 97–115.

Forthun, Ingeborg et al. (2024): "Excess mortality in Denmark, Finland, Norway and Sweden during the COVID-19 pandemic 2020–2022". *European Journal of Public Health*, ckae091, pp. 1–7.

Fraiman, Joseph et al. (2022): "Serious adverse events of special interest following mRNA COVID-19 vaccination in randomized trials in adults". *Vaccine* 40, pp. 5798–5805.

Fukuyama, Francis (1989): "The end of history?". *The National Interest* 16, pp. 3–18.
Fukuyama, Francis (1992): *The end of history and the last man*. New York, NY: Free Press.
Fukuyama, Francis (2022): *Liberalism and its discontents*. New York, NY: Farrar, Straus and Giroux.
Gates, Bill (2022): *How to prevent the next pandemic*. New York, NY: Knopf.
Gaus, Gerald F. (2003): *Contemporary theories of liberalism*. London: Sage.
Gebauer, Carlos A. (2021): "Freiheit ist mehr als gehorchen". *Eigentümlich frei*, https://ef-magazin.de/2021/11/08/19195-plandemie-narreteien-freiheit-ist-mehr-als-gehorchen, 8 November 2021.
Gerber, Kaspar (2022): *Recht und Evidenz in der Pandemie. Juristische Analysen aus zwei Jahren der Covid-19-Bekämpfung*. Zürich: EIZ Publishing.
Goklany, Indur M. (2011): *Wealth and safety: the amazing decline in deaths from extreme weather in an era of global warming*. Reason Foundation: Policy Study 393, September 2011.
Green, Toby (2021): *The Covid consensus. The new politics of global inequality*. London: Hurst.
Hadas, Edward (2021): "Seven theories of why the lockdowns happened", *Brownstone Institute*, https://brownstone.org/articles/seven-theories-of-why-the-lockdowns-happened/, 5 December 2021.
Hall, Charles A. S., Lambert, Jessica G. and Balogh, Stephen B. (2014): "EROI of different fuels and the implications for society". *Energy Policy* 64, pp. 141–152.
Hammond, John L. (2009): "Indigenous community justice in the Bolivian constitution of 2009". *Human Rights Quarterly* 33, pp. 649–681.
Hankins, James (ed). (2000): *Renaissance civic humanism: reappraisals and reflections*. Cambridge: Cambridge University Press.
Hardin, Garret (1968): "The tragedy of the commons". *Science* 162, pp. 1243–1248.
Harmange, Pauline (2020): *Moi les hommes, je les déteste*. Paris: Seuil. English translation *I hate men. Translated by Natasha Lehrer*. New York, NY: HarperCollins 2022.
Harper, Kyle (2017): *The fate of Rome: Climate, disease and the end of an empire*. Princeton, NJ: Princeton University Press.
Hart, H. L. A. (1955): "Are there any natural rights?". *Philosophical Review* 64, pp. 175–191.
Hasegawa, Tomoko et al. (2018): "Risk of increased food insecurity under stringent global climate change mitigation policy". *Nature Climate Change* 8, pp. 699–703.
Hayek, Friedrich August von (1944): *The road to serfdom*. London: Routledge.

Hayek, Friedrich August von (1952): *The counter-revolution of science. Studies on the abuse of reason.* Glencoe, IL: Free Press.

Hayek, Friedrich August von (1982): *Law, legislation and liberty.* London: Routledge.

Hayek, Friedrich August von (1976): *The denationalization of money.* London: Institute of Economic Affairs.

Herby, Jonas, Jonung, Lars and Hanke, Steve H. (2022): "A literature review and meta-analysis of the effectiveness of lockdowns on Covid-19 mortality". *Studies in Applied Economics* 200, January 2022.

Hicks, Stephen R. C. (2004): *Explaining postmodernism. Skepticism and socialism from Rousseau to Foucault.* Tempe, AZ: Scholarly Publishing.

Hoppe, Hans-Hermann (1989): *A theory of socialism and capitalism. Economics, politics, and ethics.* Dordrecht: Kluwer.

Hoppe, Hans-Hermann (2023): "On the proper study of man: reflections on method". *Quarterly Journal of Austrian Economics* 26, no. 4. https://doi.org/10.35297/qjae.010177

Hülsmann, Jörg Guido (2008): *The ethics of money production.* Auburn, AL: Ludwig von Mises Institute.

Hülsmann, Jörg Guido (2014): "Fiat money and the distribution of incomes and wealth". In D. Howden and J. Salerno (eds.): *The Fed at one hundred. A critical view on the Federal Reserve System.* Cham: Springer. Pp. 127–138.

Hülsmann, Jörg Guido (2023): *Bitcoin: its nature and its future.* 9ème Journée de Travail en Economie Autrichienne, Paris-Cachan 15 December 2023.

Hülsmann, Jörg Guido (2024): *Abundance, generosity, and the state. An inquiry into economic principles.* Auburn, AL: Ludwig von Mises Institute.

Huemer, Michael (2013): *The problem of political authority.* Houndmills: Palgrave Macmillan.

Huerta Soto, Jesus (2008): *The Austrian School: market order and entrepreneurial creativity.* Cheltenham: Edward Elgar Publishing.

Ioannidis, John P. A. (2020a): "A fiasco in the making? As the coronavirus pandemic takes hold, we are making decisions without reliable data". *First opinion*, https://www.statnews.com/2020/03/17/a-fiasco-in-the-making-as-the-coronavirus-pandemic-takes-hold-we-are-making-decisions-without-reliable-data/

Ioannidis, John P. A. (2020b): "Infection fatality rate of COVID-19 inferred from seroprevalence data". *Bulletin of the World Health Organization*, Article ID: BLT.20.265892.

IPCC (2018): *Special report: global warming of 1.5°C.* https://www.ipcc.ch/sr15/

Ivanic, Maros, Martin, Will and Zaman, Hassan (2012): "Estimating the short-run poverty impacts of the 2010-2011 surge in food prices". *World Development* 40, pp. 2302–2317.

Jackson, Alvin (2004): *Home rule: an Irish history, 1800–2000*. Oxford: Oxford University Press.

Jackson, Frank (1998): *From metaphysics to ethics*. Oxford: Oxford University Press.

Jonas, Hans (1979): *Das Prinzip Verantwortung. Versuch einer Ethik für die technologische Zivilisation*. Frankfurt (Main): Insel Verlag. English edition *The imperative of responsibility: in search of an ethics for the technological age*. Chicago, IL: University of Chicago Press 1984.

Kampf, Günter (2021): "COVID-19: stigmatising the unvaccinated is not justified". *The Lancet* 398, p. 1871.

Kant, Immanuel (1911): *Kants gesammelte Schriften. Band 4*. Edited by Königlich Preußische Akademie der Wissenschaften. Berlin: Reimer.

Kant, Immanuel (1979): *Kants gesammelte Schriften. Band 27.2.2*. Edited by Königlich Preußische Akademie der Wissenschaften. Berlin: de Gruyter.

Kant, Immanuel (1983): *Perpetual peace and other essays on politics, history, and morals. Translated by Ted Humphrey*. Indianapolis: Hackett.

Kant, Immanuel (1996): *The Cambridge edition of the works of Immanuel Kant. Volume 4. Practical philosophy. Translated and edited by Mary J. Gregor and Allen Wood*. Cambridge: Cambridge University Press.

Kant, Immanuel (2002): *The Cambridge edition of the works of Immanuel Kant. Volume 3. Theoretical philosophy after 1781. Translated and edited by Henry Allison and Peter Heath*. Cambridge: Cambridge University Press.

Karp, David R. (1999): "Community justice: six challenges". *Journal of Community Psychology* 27, pp. 751–769.

Kealey, Terence (1996): *The economic laws of scientific research*. Houndmills: Palgrave Macmillan.

Keil, Ulrich (2010): "Introductory statement". *Social, Health and Family Affairs Committee of the Parliamentary Assembly of the Council of Europe. Hearing on "The handling of the H1N1 pandemic: more transparency needed?"*, Strasbourg, 26 January 2010.

Keil, Ulrich, Schönhofer, Peter and Spelsberg, Angela (2011): "The invention of the swine-flu pandemic". *European Journal of Epidemiology* 26, pp. 187–190.

Kerpen, Phil, Moore, Stephen und Mulligan, Casey B. (2022): "A final report card on the states' response to COVID-19". *National Bureau of Economic Research Working Paper Series*, working paper 29928, April 2022, https://www.nber.org/papers/w29928

Koonin, Steven E. (2021): *Unsettled. What climate science tells, what it doesn't, and why it matters*. Dallas, TX: BenBella Books.

Kotchoubey, Boris (2012): *Why are you free? Neurobiology and psychology of voluntary action*. New York, NY: Nova Science.

Koutsoyiannis, Demetris and Vournas, Christos (2024): "Revisiting the greenhouse effect – a hydrological perspective". *Hydrological Sciences Journal* 69, pp. 151–164.

Kowall, Bernd et al. (2021): "Excess mortality due to Covid-19? A comparison of total mortality in 2020 with total mortality in 2016 to 2019 in Germany, Sweden and Spain". *PLoS ONE* 16, e0255540.

Kuhbandner, Christof, Homburg, Stefan, Walach, Harald und Hockertz, Stefan (2022): "Was Germany's lockdown in spring 2020 necessary? How bad data quality can turn a simulation into a delusion that shapes the future". *Futures* 135, e102879.

Laird, Heather (2005): *Subversive law in Ireland, 1879–1920. From 'unwritten law' to the Dáil courts*. Dublin: Four Courts Press.

Landgrebe, Jobst and Smith, Barry (2022): *Why machines will never rule the world*. London: Routledge.

Le Mouël, J. L., Lopes, F. and Courtillot, V. (2019): "Characteristic time scales of decadal to centennial changes in global surface temperatures over the past 150 years". *Earth and Space Science* 7, e2019EA000671.

Leeson, Peter T. (2014): *Anarchy unbound: why self-governance works better than you think*. Cambridge: Cambridge University Press.

Leoni, Bruno (1961): *Freedom and the law*. Princeton, NJ: D. Van Nostrand.

Leopoldina (2020a): "7. Ad-hoc-Stellungnahme zur Coronavirus-Pandemie", https://www.leopoldina.org/publikationen/detailansicht/publication/coronavirus-pandemie-die-feiertage-und-den-jahreswechsel-fuer-einen-harten-lockdown-nutzen-2020/, 8 December 2020.

Leopoldina (2020b): "Mission Statement of the German National Academy of Sciences Leopoldina", https://www.leopoldina.org/en/about-us/about-the-leopoldina/leopoldina-mission-statement/

Leplin, Jarrett (1987): "The bearing of discovery on justification". *Canadian Journal of Philosophy* 17, pp. 805–814.

Levine, Joseph (1983): "Materialism and qualia: The explanatory gap". *Pacific Philosophical Quarterly* 64, pp. 354–361.

Lewis, David (1966): "An argument for the identity theory". *Journal of Philosophy* 63, pp. 17–25.

Lindzen, Richard S. (2020a): "An oversimplified picture of the climate behavior based on a single process can lead to distorted conclusions". *European Physical Journal Plus* 135, article no. 462.

Lindzen, Richard S. (2020b): "On climate sensitivity". http://co2coalition.org/wp-content/uploads/2020/04/Climate-Sensitivity-06.07.20.pdf

Lindzen, Richard S. and Christy, John R. (2020): "The global mean temperature anomaly record. How it works and why it is misleading". https://co2coalition.org/publications/the-global-mean-temperature-anomaly-record/

Locke, John (1823): *Two treatises of government*. In: *The works of John Locke. A new edition, corrected. In ten volumes. Volume V*. London: Tegg.

Lottieri, Carlo (2002): "European unification as the new frontier of collectivism: the case for competitive federalism and polycentric law". *Journal of Libertarian Studies* 16, pp. 23–43.

Lyotard, Jean-François (1979): *La condition postmoderne*. Paris: Minuit. English *The postmodern condition. Translated by Geoff Bennington and Brian Massumi*. Minneapolis, MA: University of Minnesota Press 1984.

Mader, Sebastian und Rüttenauer, Tobias (2022): "The effects of non-pharmaceutical interventions on COVID-19 mortality: a generalized synthetic control approach across 169 countries". *Frontiers in Public Health* 10, e820642.

Maiocchi, Roberto (1999): *Scienza italiana e razzismo fascista*. Firenze: La Nuova Italia.

Manifeste conspirationniste (2022). Paris: Seuil.

Marx, Karl and Engels, Friedrich (1962): *Werke. Band 19*. Berlin (Ost): Dietz.

Marx, Karl and Engels, Frederick (1966): *Selected works. Volume III*. Moscow: Progress Publishers.

Masselot, Pierre et al. (2023): "Excess mortality attributed to heat and cold: a health impact assessment study in 854 cities in Europe". *Lancet Planetary Health* 7, e271–281.

Massimi, Michela (2022): *Perspectival realism*. Oxford: Oxford University Press.

Maycock, Amanda C. et al. (2018): "Revisiting the mystery of recent stratospheric temperature trends". *Geophysical Research Letters* 45, pp. 9919–9933.

McDowell, John (1994): *Mind and world*. Cambridge, MA: Harvard University Press.

McDowell, John (1995): "Two sorts of naturalism". In: R. Hursthouse, G. Lawrence and W. Quinn (eds): *Virtues and reasons: Philippa Foot and moral theory*. Oxford: Oxford University Press. Pp. 149–179.

McKitrick, Ross and Christy, John R. (2020): "Pervasive warming bias in CMIP6 tropospheric layers". *Earth and Space Science* 7, e2020EA001281.

Mele, Alfred R. (2014): *Free: why science hasn't disproved free will*. Oxford: Oxford University Press.

Merkel, Angela (2020): "Rede von Bundeskanzlerin Dr. Angela Merkel zum Haushaltsgesetz 2021 vor dem Deutschen Bundestag am 9. Dezember 2020 in Berlin", https://www.bundesregierung.de/breg-de/service/bulletin/rede-von-bundeskanzlerin-dr-angela-merkel-1826624, 9 December 2020.

Merton, Robert K. (1942): "Science and technology in a democratic order". *Journal of Legal and Political Sociology* 1, pp. 115–126.

Michelson, Albert A. (1894): "Dedication of Ryerson Physical Laboratory". In: *University of Chicago Annual Register July 1895–July 1896*. Chicago, IL: University of Chicago Press. P. 159.

Michelson, Albert A. (1903): *Light waves and their uses*. Chicago, IL: University of Chicago Press.

Miles, David K., Stedman, Michael und Heald, Adrian H. (2020): "'Stay at home, protect the National Health Service, safe lives': a cost benefit analysis of the lockdown in the United Kingdom". *International Journal of Clinical Practice* 75.3, DOI: https://doi.org/10.1111/ijcp.13674.

Mill, John Stuart (1885): *Principles of political economy*. New York, NY: Appleton. Original publication 1848.

Mises, Ludwig von (1949): *Human action*. New Haven, CT: Yale University Press. Reprint Auburn, Al: Ludwig von Mises Institute 1998.

Mullis, Kary et al. (1997): "Corporate greed and AIDS". Conference, Santa Monica, California, 12 July 1997. Statements on PCR tests on https://www.youtube.com/watch?v=wT3IqZjT_9A

Nagel, Thomas (1986): *The view from nowhere*. Oxford: Oxford University Press.

Nebehay, Stephanie (2008): "Top U.N. human rights forum to examine food crisis". https://www.reuters.com/article/idUSL09119643

Niemietz, Kristian (2019): *Socialism. The failed idea that never dies*. London: The Institute of Economic Affairs.

Nozick, Robert (1974): *Anarchy, state, and utopia*. New York, NY: Basic Books.

Othman, Anwar Hasan Abdullah, Syed Musa Alhabshi, Salina Kassim, Adam Abdullah and Razali Haron (2020): "The impact of monetary systems on income inequity and wealth distribution". *International Journal of Emerging Markets* 15, pp. 1161–1183.

Patton, Paul (2001): "Postmodernism: philosophical aspects". In N. J. Smelser and P. B. Baltes (eds.): *International encyclopedia of the social and behavioral sciences*. Amsterdam: Elsevier. Pp. 11872–11877.

Peels, Rik (2023): "Scientism and scientific fundamentalism: what science can learn from mainstream religion". *Interdisciplinary Science Reviews* 48, pp. 395–410.

Pezzullo, Angelo Maria et al. (2023): "Age-stratified infection fatality rate of COVID-19 in the non-elderly population". *Environmental Research* 216, 114655.

Pfizer Documents Investigations Team (2023): *Pfizer documents analysis reports*. DailyClout.

Planck, Max (1924): "Vom Relativen zum Absoluten". Reprinted in Max Planck (2001): *Vorträge, Reden, Erinnerungen. Herausgegeben von Hans Roos und Armin Hermann*. Berlin: Springer. Pp. 103–117.

Polleit, Thorsten (2022a): *The global currency plot: how the deep state will betray your freedom, and how to prevent it*. Auburn (Alabama): Mises Institute.

Polleit, Thorsten (2022b): *Der Weg zur Wahrheit. Eine Kritik der ökonomischen Vernunft*. München: FinanzBuch Verlag.

Popper, Karl (1935): *Logik der Forschung*. Wien: Springer. English *The logic of scientific discovery*. London: Hutchinson 1959.
Popper, Karl (1945): *The open society and its enemies*. London: Routledge.
Popper, Karl (1957): *The poverty of historicism*. London: Routledge.
Popper, Karl (1997): "Tribute to the life and work of Friedrich Hayek". In S. F. Frowen (ed.): *Hayek: economist and social philosopher. A critical retrospect*. Houndmills: Macmillan. Pp. 311–312.
Prugger, Christof et al. (2021): "Evaluating covid-19 vaccine efficacy and safety in the post-authorisation phase". *British Medical Journal* 375, e067570.
Psillos, Stathis (1999): *Scientific realism. How science tracks truth*. London: Routledge.
Putnam, Hilary (1975): "The nature of mental states". In: H. Putnam: *Mind, language and reality. Philosophical papers volume 2*. Cambridge: Cambridge University Press. Pp. 429–440.
Raffelhüschen, Bernd (2020): "Verhältnismässigkeit in der Pandemie: Geht das?". *WiSt. Wirtschaftswissenschaftliches Studium*, July 2020.
Rand, Ayn (1964): *The virtue of selfishness*. New York, NY: New American Library.
Rawls, John (2001): *Justice as fairness. A restatement. Edited by Erin I. Kelly*. Cambridge, MA: Harvard University Press.
Rectenwald, Michael (2018): *Springtime for snowflakes: "social justice" and its postmodern parentage*. Nashville, TN: New English Review Press.
Rectenwald, Michael (2021): "Social justice and the emergence of Covid tyranny". *Mises Institute*, https://mises.org/print/56271, 14 September 2021.
Rectenwald, Michael (2023): *The great reset and the struggle for liberty. Unraveling the global agenda*. Nashville, TN: New English Review Press.
Richet, Pascal (2021): "The temperature–CO_2 climate connection: an epistemological reappraisal of ice-core messages". *History of Geo- and Space Sciences* 12, pp. 97–110.
Riemersma, Kasen K. et al. (2022): "Shedding of infectious SARS-CoV-2 despite vaccination". *PLOS Pathogens*, DOI https://doi.org/10.1371/journal.ppat.1010876, 30 September 2022.
Rorty, Richard (1980): *Philosophy and the mirror of nature*. Oxford: Blackwell.
Rorty, Richard (1999): *Philosophy and social hope*. London: Penguin.
Rothbard, Murray N. (1982): *The ethics of liberty*. New York, NY: New York University Press.
Ruddiman, William F. (2014): *Earth's climate: past and future*. New York, NY: Freeman. Third edition.
Saint Simon, Claude-Henri de (1865–1878): *Œuvres de Saint-Simon & d'Enfantin. XXVII volumes*. Paris: Dentu.
Schickore, Jutta (2014): "Scientific discovery". In E. N. Zalta (ed.): *The Stanford Encyclopedia of Philosophy*, https://plato.stanford.edu/entries/scientific-discovery/

Schneider, Stephen H. (1989): "The greenhouse effect: science and policy". *Science* 243, pp. 771–781.
Schwab, Klaus and Malleret, Thierry (2020): *COVID-19: the great reset.* Geneva: World Economic Forum.
Schwab, Klaus and Malleret, Thierry (2022): *The great narrative. For a better future.* Geneva: World Economic Forum.
Sellars, Wilfrid (1956): "Empiricism and the philosophy of mind". In: H. Feigl and M. Scriven (eds.): *The foundations of science and the concepts of psychology and psychoanalysis.* Minneapolis: University of Minnesota Press. Pp. 253–329.
Sellars, Wilfrid (1962): "Philosophy and the scientific image of man". In: R. Colodny (ed.): *Frontiers of science and philosophy.* Pittsburgh, PA: University of Pittsburgh Press. Pp. 35–78.
Shellenberger, Michael (2020): *Apocalypse never. Why environmental alarmism hurts us all.* New York, NY: Harper Collins.
Simmons, A. John (1979): *Moral principles and political obligations.* Princeton, NJ: Princeton University Press.
Sinder, Janet (2001): "Irish legal history: an overview and guide to the sources". *Law Library Journal* 93, pp. 231–260.
Singanayagam, Anika et al. (2022): "Community transmission and viral load kinetics of the SARS-CoV-2 delta (B.1.617.2) variant in vaccinated and unvaccinated individuals in the UK: a prospective, longitudinal, cohort study". *Lancet Infectious Diseases* 22, pp. 183–195.
Smil, Vaclav (2016): *Power density: a key to understanding energy sources and uses.* Cambridge, MA: MIT Press.
Smith, Adam (1976): *An inquiry into the nature and causes of the wealth of nations.* Oxford: Oxford University Press. Original publication 1776.
Sowell, Thomas (1995): *The vision of the anointed. Self-congratulation as a basis for social policy.* New York, NY: Basic Books.
Stang, Andreas et al. (2021): "The performance of the SARS-CoV-2-PCR test as a tool for detecting SARS-CoV-2 infection in the population". *Journal of Infection* 83, pp. 237–239.
Steiner, Rudolf (1966): *The threefold social order. Translated by Frederick C. Heckel.* New York, NY: Anthroposophic Press.
Steiner, Rudolf (1985): *The renewal of the social organism. Translated by E. Bowen-Wedgewood and Ruth Mariott.* New York, NY: Anthroposophic Press.
Stepan, Nancy Leys (1992): "Race, gender and nation in Argentina: the influence of Italian eugenics". *History of European Ideas* 15, pp. 749–756.
Streeck, Hendrik et al. (2020): "Infection fatality rate of SARS-CoV2 in a super-spreading event in Germany". *Nature Communications* 11, 5829.
Subramanian, S. V. und Kumar, Akhil (2021): "Increases in COVID-19 are unrelated to levels of vaccination across 68 countries and 2947 counties in the United States". *European Journal of Epidemiology* 36, pp. 1237–1240.

Templeton, Steve (2023): *Fear of a microbial planet: how a germophobic safety culture makes us less safe*. Austin, TX: Brownstone Institute.

Thakur, Ramesh (2023): *Our enemy, the government: how Covid enabled the expansion and abuse of state power*. Austin, TX: Brownstone Institute.

Thomson, Judith Jarvis (1990): *The realm of rights*. Cambridge, MA: Harvard University Press.

Tiedtke, Andreas (2021): *Der Kompass zum lebendigen Leben*. München: FinanzBuch Verlag.

Tierney, Brian (1982): *Religion, law, and the growth of constitutional thought 1150–1650*. Cambridge: Cambridge University Press.

Tol, Richard S. J. (2016): "Comment on 'Quantifying the consensus on anthropogenic global warming in the scientific literature'". *Environmental Research Letters* 11, 048001.

Tucker, Jeffrey (2019): *The market loves you: why you should love it back*. Great Barrington, MA: American Institute of Economic Research.

United Nations (2020): "Economic and Social Council. Committee on Economic, Social and Cultural Rights. General comment No. 25 (2020) on science and economic, social and cultural rights (article 15 (1) (b), (2), (3) and (4) of the International Covenant on Economic, Social and Cultural Rights)", https://docstore.ohchr.org/SelfServices/FilesHandler.ashx?enc=4slQ6QSmlBEDzFEovLCuW1a0Szab0oXTdImnsJZZVQdxONLLLJiul8wRmVtR5Kxx73i0Uz0k13FeZiqChAWHKFuBqp%2B4RaxfUzqSAfyZYAR%2Fq7sqC7AHRa48PPRRALHB

Vasquez, Ian, McMahon, Fred, Murphy, Ryan and Sutter Schneider, Guillermina (2023): *The human freedom index 2023*. Washington, DC: Cato Institute and Vancouver: Fraser Institute.

Verano, Alfredo (1929): *Las doctrinas eugénicas (ensayo de sistematización)*. Buenos Aires: edición del autor.

Villa-Vicencio, Charles (2000): "Christianity and human rights". *Journal of Law and Religion* 14, pp. 579–600.

Walzer, Uwe and Hendel, Roland (2023): "Natural climate change and glaciations". *Earth-Science Reviews* 241, 104435.

Watson, James D. and Crick, Francis H. C. (1953): "A structure for deoxyribose nucleic acid". *Nature* 171, pp. 737–738.

Weber, Max (1919): *Politik als Beruf. Geistige Arbeit als Beruf. Vier Vorträge vor dem Freistudentischen Bund. Zweiter Vortrag. München: Duncker & Humblot.* Most recent English translation in *Weber's rationalism and modern society. New translations on politics, bureaucracy, and social stratification. Edited and translated by Tony Waters and Dagmar Waters.* New York, NY: Palgrave Macmillan 2015.

Weber, Max (1922): *Grundriss der Sozialökonomik. III. Abteilung. Wirtschaft und Gesellschaft*. Tübingen: Mohr. English translation *The theory of social and eco-*

nomic organization. Translated by A. M. Henderson and Talcott Parsons. Oxford: Oxford University Press 1947.

Weißbach, Daniel et al. (2013): "Energy intensities, EROIs (energy returned on invested), and energy payback times of electricity generating power plants". *Energy* 52, pp. 210–221.

Wittgenstein, Ludwig (1953): *Philosophical investigations. Translated by G. E. M. Anscombe.* Oxford: Blackwell.

World Health Organization (2019): "Global influenza programme: Non-pharmaceutical public health measures for mitigating the risk and impact of epidemic and pandemic influenza". https://apps.who.int/iris/bitstream/handle/10665/329438/9789241516839-eng.pdf

Index[1]

[1] Note: Page numbers followed by 'n' refer to notes.

M. Esfeld, C. Lopez, *Restoring Science and the Rule of Law*, Palgrave Studies in Austrian Economics,
https://doi.org/10.1007/978-3-031-71186-2

SPRINGER NATURE

GPSR Compliance

The European Union's (EU) General Product Safety Regulation (GPSR) is a set of rules that requires consumer products to be safe and our obligations to ensure this.

If you have any concerns about our products, you can contact us on ProductSafety@springernature.com

In case Publisher is established outside the EU, the EU authorized representative is:

Springer Nature Customer Service Center GmbH
Europaplatz 3
69115 Heidelberg, Germany

The manufacturer's authorised representative in the EU is Springer Nature Customer Service Centre GmbH, Europaplatz 3, 69115 Heidelberg, Germany. If you have any concerns regarding our products, please contact ProductSafety@springernature.com

Printed and bound by CPI Group (UK) Ltd, Croydon, CR0 4YY

01/10/2026

02225903-0005